Contents

Preface   1

Part 1
The position of humans in the animal kingdom

1 Humans and other animals   2

Part 2
Cells and metabolism

2 Cells: the units of life   4
3 Cell structure   6
4 Metabolism and enzymes   8
5 Respiration   10
6 Adenosine triphosphate (A.T.P.)   12

Part 3
Tissues, organs, and organ systems

7 Epithelial tissues   14
8 Muscle tissues   16
9 Nervous tissues and nerve impulses   18
10 Connective tissues   20
11 Organs and organ systems   22

Part 4
The skeleton, muscles, and movement

12 The skeleton and its functions   24
13 Structure and growth of bones   26
14 Joints   28
15 Muscles, movement, and leverage   30
16 Sprains, dislocations, and fractures   32
17 Posture   34
18 Exercise, fatigue, and rest   36

Part 5
Nutrition

19 Photosynthesis and food chains   38
20 Types of food and their energy value   40
21 Vitamins and minerals   42
22 Diet   44
23 Food preservation   46
24 Digestion, absorption, and assimilation   48
25 Teeth   50
26 Digestion   52
27 Absorption   54
28 The liver   56

Part 6
The circulatory system

29 Blood and its functions   58
30 Structure of the heart   60
31 How the heart pumps blood   62
32 Arteries, capillaries, and veins   64
33 The main blood vessels   66
34 Tissue fluid and lymph   68
35 Blood groups and blood transfusions   70
36 Blood clotting, and treatment of bleeding and wounds   72

Part 7
The respiratory system

37 Structure of the respiratory system   74
38 Breathing movements   76
39 Lung volume, breathing rate, and artificial respiration   78
40 Gaseous exchange   80

Part 8
Homeostasis

41 Homeostasis, excretion, and the urinary system   82
42 Excretion by the kidneys   84
43 Skin and temperature control   86
44 Structure and care of skin and hair   88

Part 9
Sense organs

45 Senses in the skin, tongue, and nose   90
46 Structure of the eye   92
47 Focusing and structure of the retina   94
48 Binocular vision and eye defects   96
49 The ear and hearing   98
50 Sense of balance   100

Part 10
The nervous system

51 Parts of the nervous system   102
52 Reflexes   104
53 The autonomic nervous system   106
54 The endocrine system   108

Part 11
Reproduction

55 Male reproductive system   110
56 Female reproductive system   112
57 The menstrual cycle   114
58 Fertilization and implantation   116
59 Pregnancy and the placenta   118
60 Development before birth   120
61 Birth   122
62 The newborn baby   124
63 Development of co-ordination   126
64 Growth, development, and ageing   128
65 Birth control   130

Part 12
Heredity

66 Variation, heredity, and genetics   132
67 Chromosomes, genes, and mitosis   134
68 Gametes, meiosis, and sex chromosomes   136
69 Heredity and genes   138
70 Some technical terms used in genetics   140
71 DNA and the genetic code   142
72 Transcribing the genetic code   144

Part 13
The causes and prevention of disease

73 Some causes of disease   146
74 Some microbes which cause disease   148

75 Parasitic roundworms (nematodes)   150
76 Tapeworms   152
77 Fleas, lice, and bed bugs   154
78 Houseflies and disease   156
79 The spread of infection   158
80 Prevention of infection   160
81 Pure water supplies   162
82 Disposal of sewage and refuse   164
83 Safety in and around the home   166
84 Ventilation, heating, and lighting   168
85 Pollution   170
86 Smoking   172
87 Drinking and drug taking   174
88 The National Health Service   176
89 Antiseptic surgery and medical drugs   178

Part 14
Immunity

90 Natural immunity   180
91 Acquired immunity and immunization   182

Revision tests   184

Glossary   199

Index   208

Acknowledgements

The publishers would like to thank the following for permission to reproduce photographs:

Camera Press, pp. 171 (bottom right), 171 (top right), 175 (top right); J Allan Cash, p. 171 (centre left); Crown Copyright, p. 71; Health Education Council, pp. 172, 175 (bottom); Hills Harris, pp. 35, 36–7; J G Hirsch, p. 180; Oxfam, p. 141; Picturepoint Ltd., pp. 171 (bottom left), 171 (centre right), 171 (top left); St Mary's Hospital Medical School, p. 179 (bottom left); Seaphot Ltd., p. 79; Chris Schwarz, p. 175 (top left); University College Hospital (L D Ellis), p. 179 (bottom right); James Webb, pp. 6, 134;

By courtesy of the Wellcome Trustees, p. 179 (top right); Terry Williams, p. 131.

The line drawing on pp. 162–3 was adapted and reproduced with the kind permission of the Yorkshire Water Authority.

The drawings on the following pages were done by Brian and Constance Dear: 2, 3, 8, 9, 22, 23, 24, 31 (bottom), 32, 33, 34, 35, 40, 43, 44, 45, 47, 48, 49, 69 (bottom), 71, 73 (bottom), 79, 103 (bottom), 107, 108, 125, 126 (left), 127, 129, 132, 133, 147, 159, 160, 161, 166, 167, 177.

Preface

This book covers the main topics required by GCSE syllabuses in Human Biology. It will also supplement GCSE Biology courses.

The book contains 91 double-page units, each of which presents a self-contained explanation of a particular topic, but with cross-references to other relevant units. Text is confined to the left-hand page of each unit. The remaining space (at least one page) is devoted to drawings, diagrams, charts, or photographs. This arrangement enables a student to study a topic in short, easily assimilated steps, with the aid of far more illustrations than are found in more conventional textbooks.

The book has two types of illustrations. There are realistic drawings, designed to attract attention and stimulate interest, which contain sufficient detail to give a clear impression of specimens as they appear in nature. These drawings can be especially useful where specimens are not available, as in the case of home study. In addition there are simple diagrams designed to show how realistic drawings can be reduced to their essential elements, or which provide a diagrammatic summary of the biological principles covered by the unit.

The text is written in simple language. Technical terms are fully explained and printed in bold type when they first appear, so that they can easily be revised in context. In addition a glossary is provided to help students learn technical terms.

At the end of the book there is a section of revision tests. These tests are of two kinds: vocabulary tests, which enable students to check their knowledge of technical terms used in the text; and comprehension tests, which establish whether they have understood what they have read. In certain units there are illustrations which take the form of a quiz to test understanding of the text. The author's own classroom experience has shown that most students find these three types of test interesting to complete while teachers find them easy to administer and mark.

Humans and other animals

Humans are members of the animal kingdom. They belong to a group of animals called **mammals**, which also includes kangaroos, dolphins, horses, and bats, to name just a few. All mammals have a constant warm body temperature (they are warm-blooded) and have a hairy skin. Baby mammals develop inside their mother's body and after birth are fed on milk from her mammary glands (breasts).

Humans belong to a group of mammals called **anthropoids**. The other anthropoids are monkeys, gibbons, and apes. Anthropoids have the following characteristics. They have large brains and are the most intelligent of all animals. Their behaviour is inquisitive and adventurous and they learn quickly. The human brain is more than twice as large as that of any other anthropoid and gives humans their great powers of learning and problem solving (Fig. 3).

The eyes of an anthropoid face forwards and not sideways as in many other animals (Fig. 2). This is important because it gives them a good judgement of distance. The face is rather flat, especially in humans, and the ears are small and lie flat against the skull. In humans the lower jaw forms a chin (Figs. 1 and 2).

Anthropoids can stand upright although, except for humans, they usually walk on all fours. Humans have shorter arms than other anthropoids (Fig. 1) and their bodies are comparatively hairless. The hands of all anthropoids are capable of grasping objects. They are used to pick up and eat food and can manipulate objects for use as tools. All anthropoids except humans have big toes which can be used like a thumb (Fig. 4).

Anthropoids can 'pull faces'. They can move their mouths, eyes, cheeks, and noses to make many different facial expressions. These expressions, together with body movements and a great variety of sounds, allow them to communicate with one another. Humans make complex sounds called speech.

Anthropoids are social animals. They live in groups which usually have a leader, and individuals help look after the group as a whole as well as themselves. Human social groups form for reasons other than survival. People can belong to several groups according to their age, sex, occupation, income, nationality, race, interests, etc.

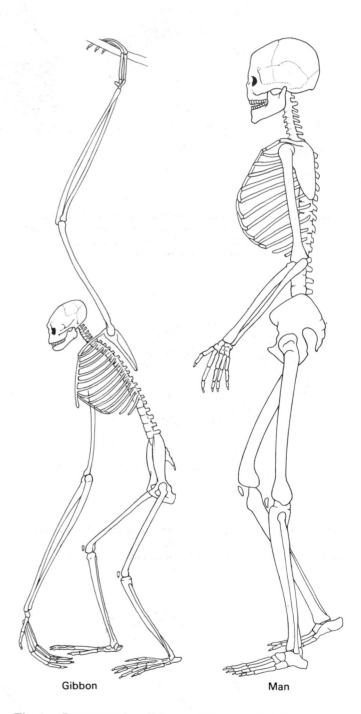

Gibbon   Man

Fig. 1. Compare the gibbon and human skeletons.
Humans' legs are long and their arms are short compared with an ape. Humans are bipedal (balance on two legs) whereas apes prop up the front of the body with their long arms. The human skull has a volume of 1500 cm³. It has a high forehead, no eyebrow ridges, and a prominent chin. The largest ape skull has a volume of only 650 cm³. It has a receding forehead and chin and prominent eyebrow ridges.

Fig. 2.   Compare these three faces. The horse has a long snout, eyes on the sides of its head, and large ears which stick up from its head. How is the gorilla's face different from the horse's, and how is the man's face different from the gorilla's?

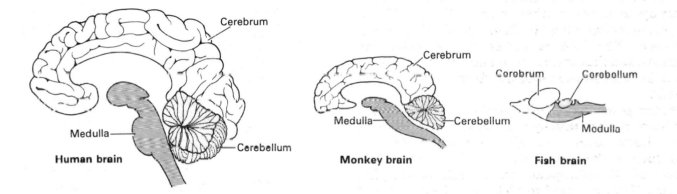

Fig. 3.   Cross-sections through three brains. Compare the size of the human cerebrum, cerebellum, and medulla with the other two. Note also the folded cerebrum in humans and monkeys.

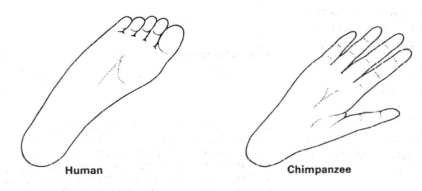

Fig. 4.   Compare the human and chimpanzee foot. How is one better adapted for grasping than the other?

2

Cells: the units of life

The bodies of animals and plants are not made of continuous material. They are made up of separate units called cells. The human body, for example, consists of millions of cells but most are too small to see unless highly magnified. Figure 2 compares the size of a few cells with the thickness of a human hair.

The word **multicellular** is used to describe organisms which consist of many cells. All animals and plants are multicellular. The word **unicellular** can be used to describe tiny creatures like bacteria, *Amoeba*, and *Paramecium*, since they consist of only one cell (Fig. 3).

Characteristics of plant and animal cells

The detailed structure of cells is described in the next unit. Here it is sufficient to say that all cells consist of a very complex living material called **protoplasm**, and are made up of three basic parts. They have a thin skin called the **cell membrane**. This membrane encloses a jelly-like substance called **cytoplasm** and a small object called a **nucleus** (Fig. 1A).

Plant cells All plant cells are enclosed in a layer of tough material called **cellulose**. This layer is called the cell wall and it is on the outside of the cell membrane (Fig. 1B). Fully formed plant cells contain one or more large, permanent spaces called **vacuoles**. These spaces are filled with a liquid called **cell sap** and are lined with a skin similar to the cell membrane. The green parts of a plant are green because their cells contain **chlorophyll**. This is a green pigment and it is contained inside tiny objects called **chloroplasts** which float in the cytoplasm. Chlorophyll absorbs the light energy which plants use to combine carbon dioxide gas and water to make sugar for their food. This process is called **photosynthesis**.

Animal cells Animal cells have no cellulose wall, and never contain chlorophyll. Small temporary vacuoles often occur in their cytoplasm (Fig. 1A). There is a greater variety of shape and function amongst animal cells than amongst plant cells.

A   A diagram of an animal cell

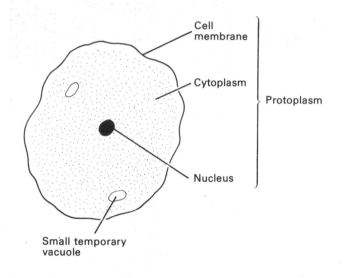

B   A diagram of a plant cell

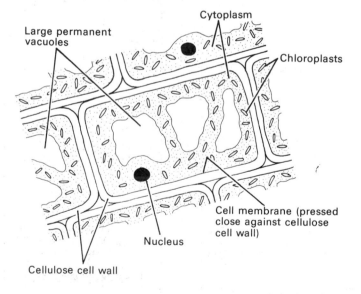

Fig. 1.   Animals and plants are made up of units called cells. All cells are made of living material called protoplasm. Cells have three main parts: a cell membrane, jelly-like cytoplasm, and a nucleus. Plant cells are enclosed in a cellulose wall and their cytoplasm contains fluid-filled spaces called vacuoles and green chloroplasts. Chloroplasts contain the chlorophyll which traps the light energy plants need for photosynthesis. Animal cells do not have a cellulose wall, never contain chlorophyll, and have only small temporary vacuoles.

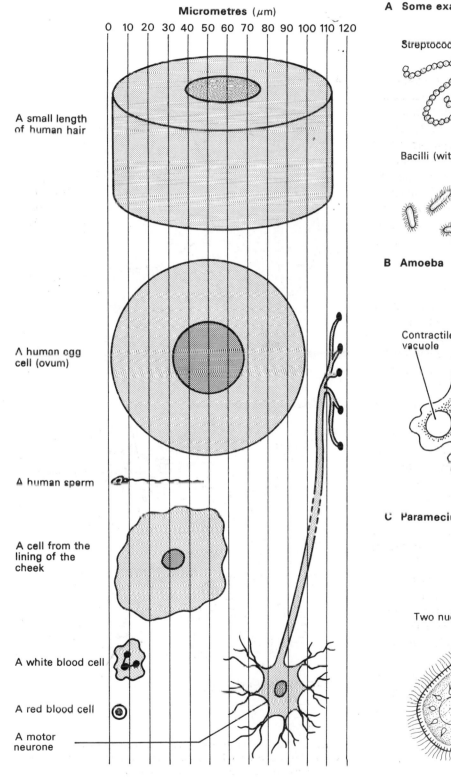

Micrometres (μm)

0 10 20 30 40 50 60 70 80 90 100 110 120

A small length of human hair

A human egg cell (ovum)

A human sperm

A cell from the lining of the cheek

A white blood cell

A red blood cell

A motor neurone

Fig. 2. Compare the size of the cells with the thickness of a human hair. These objects are drawn 550 times larger than life.

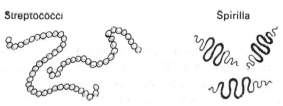

A Some examples of bacteria (× 1000)

Streptococci

Spirilla

Bacilli (with and without flagella)

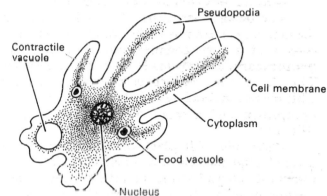

B Amoeba (× 60) (lives in ponds and streams)

Pseudopodia

Contractile vacuole

Cell membrane

Cytoplasm

Food vacuole

Nucleus

C Paramecium (× 300) (lives in ponds and streams)

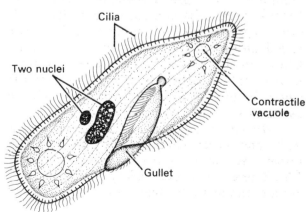

Cilia

Two nuclei

Contractile vacuole

Gullet

Fig. 3. These creatures consist of one cell and so are known as unicellular organisms. Multicellular organisms (all animals and plants) consist of many cells.

3
Cell structure

All cells are enclosed in an extremely thin 'skin' called the **cell membrane**. This membrane allows certain substances to move in and out of cells but prevents the movement of others. Consequently it is said to be **semi-permeable**.

The bulk of a cell is made up of a jelly-like substance called **cytoplasm**. Cytoplasm contains tubular and slit-like passageways called the **endoplasmic reticulum**. This reticulum probably allows materials to move quickly throughout the cell. Cytoplasm also contains starch grains, glycogen granules, oil droplets, and crystals of substances to be excreted. These non-living objects are called **inclusions**.

In addition, there are many living objects in cytoplasm called **organelles**. All cells contain round or sausage-shaped organelles called **mitochondria** (*singular* mitochondrion). These are the power plants of a cell. Mitochondria contain enzymes which release energy from food. This process is called respiration and is described in Unit 5.

Tiny organelles called **ribosomes** float in the cytoplasm and are attached to the endoplasmic reticulum. Ribosomes contain a chemical called **ribonucleic acid** (**RNA**). They take part in the manufacture of proteins, described in Unit 72.

Organelles called **lysosomes** contain enzymes which are released when cells are injured. Lysosomes digest parts of cells and ingested food.

Cytoplasm contains a number of flattened spaces called the **Golgi body**. This is thought to store and concentrate useful substances before they are removed (secreted) from the cell.

All cells contain at least one large spherical or oval object called a **nucleus**. It has a membrane perforated with holes through which chemicals move to and from the cytoplasm, and it is rich in a chemical called **deoxyribonucleic acid** (**DNA**). This chemical forms part of objects called **chromosomes**, which only become visible in a nucleus when a cell divides. DNA is concerned with protein manufacture, and plays an essential part in cell division, a process described in Units 67 and 68.

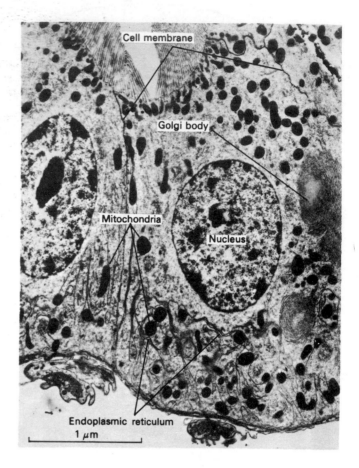

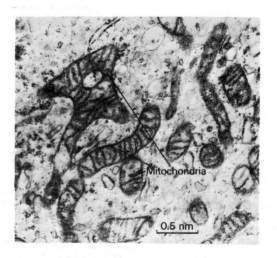

Electron micrographs of cells.
A: A cell from the intestine wall ($1 \mu m = 1/1000$ mm)
B: Mitochondria (1 nm $= 1/1000 \mu m$)

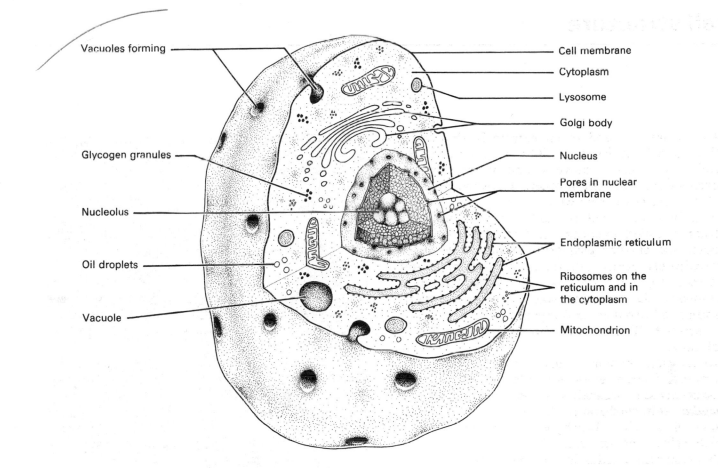

Vacuoles forming

Cell membrane

Cytoplasm

Lysosome

Golgi body

Glycogen granules

Nucleus

Pores in nuclear membrane

Nucleolus

Oil droplets

Endoplasmic reticulum

Ribosomes on the reticulum and in the cytoplasm

Vacuole

Mitochondrion

Fig. 1.   A greatly magnified animal cell cut open to show its internal structure.

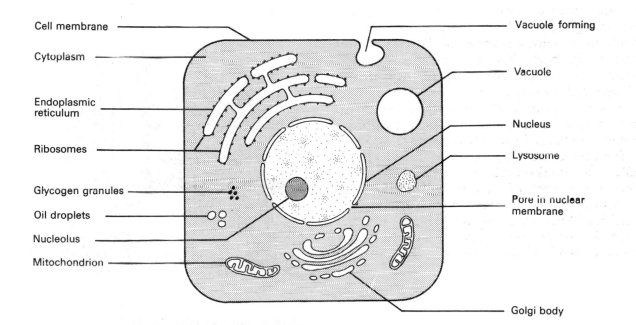

Cell membrane

Vacuole forming

Cytoplasm

Vacuole

Endoplasmic reticulum

Nucleus

Ribosomes

Lysosome

Glycogen granules

Pore in nuclear membrane

Oil droplets

Nucleolus

Mitochondrion

Golgi body

Fig. 2.   Diagram of the main parts of an animal cell.

7

4

Metabolism and enzymes

Metabolism is the technical name for all the chemical changes in an organism which are necessary for life. Metabolism involves more than a hundred different processes, but each of these can be grouped under one of two headings (Fig. 1).

1. There are a number of chemical changes in which large molecules are broken down into smaller ones with the release of energy. Reactions of this type are examples of **catabolism**. During respiration, for example, glucose sugar is broken down into carbon dioxide gas and water releasing energy which keeps the body alive.

2. Another different set of chemical changes make use of the energy released during catabolism to build (synthesize) large molecules from smaller ones. Reactions of this type are examples of **anabolism**. Figure 2 shows glucose molecules being used to build large starch molecules.

Enzymes

Metabolism would occur very slowly or not at all if it were not for chemicals called enzymes. Enzymes speed up chemical changes inside organisms without themselves being used up in the change. The word **catalyst** is used to describe chemicals which do this.

Figures 2 and 3 illustrate the way in which enzymes are thought to work. An enzyme combines briefly with a substance and activates it so that it undergoes a chemical change. The changed chemicals are then released from the enzyme which is immediately ready for another reaction.

Each type of enzyme acts as a catalyst for only one type of chemical change. In other words, enzymes are *specific*. Proteases, for example, are enzymes which break down (digest) only protein foods and nothing else (Fig. 3).

Temperature influences the speed at which enzymes work. A rise in temperature causes an increase in the speed of a reaction up to the enzyme's optimum temperature. Above this point the reaction goes faster for a time but soon the enzyme is destroyed by the heat. Some enzymes work best in alkaline conditions, others in acids, and others where conditions are neutral. In other words, each enzyme requires a specific pH level for optimum efficiency.

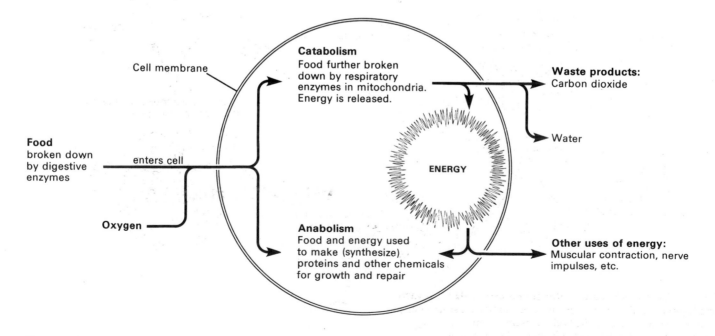

Fig. 1. Diagram of metabolism: all the chemical changes which are necessary for life.

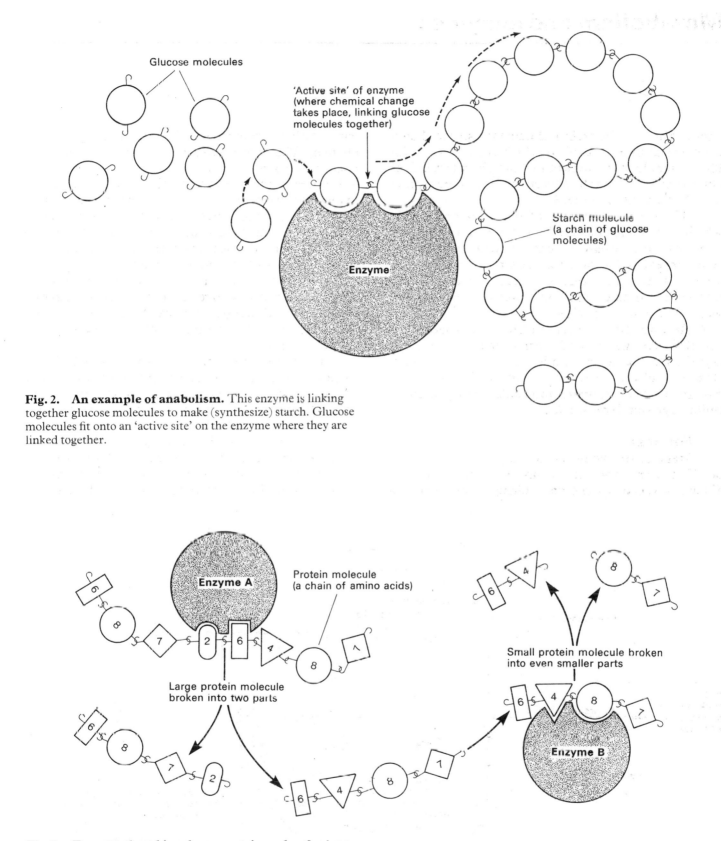

Fig. 2.   An example of anabolism. This enzyme is linking together glucose molecules to make (synthesize) starch. Glucose molecules fit onto an 'active site' on the enzyme where they are linked together.

Fig. 3.   Enzymes breaking down protein molecules into small parts. This is not an example of catabolism since energy is not released for use by the body. It is an example of how digestive enzymes work, and also shows that enzymes are specific: enzyme A will only split apart amino acids 2 and 6, and enzyme B only splits apart acids 4 and 8.

5

Respiration

Respiration is the chemical breakdown of foods such as sugars and fats to produce energy for life. This process is often called **cellular** or **internal respiration** because it takes place inside body cells.

Cells release all the available energy from food by breaking it down into carbon dioxide and water. To do this cells require a supply of oxygen. For example, the complete breakdown of one gram molecule of glucose sugar (180 g) can be summarized as follows:

$$C_6H_{12}O_6 \ + \ 6O_2 \ \xrightarrow{\text{Enzymes}} \ 6CO_2 \ + \ 6H_2O \ + \ 2898\,kJ$$

glucose   *oxygen*   *carbon*   *water*   *energy*
dioxide

This equation shows only the raw materials and end-products of respiration. The overall process involves some fifty different reactions, each catalysed by its own enzyme. Figure 1 illustrates some of the many ways in which energy released during respiration is used in the body.

The fifty-odd reactions of respiration occur in two stages. The first stage does not require oxygen and releases very little energy because food is only partly broken down: it is reduced to intermediate compounds such as alcohol or lactic acid. This first stage in which oxygen is *not* required is called **anaerobic respiration**. The second stage of respiration does require oxygen. It is called **aerobic respiration**, and involves the complete breakdown of the intermediate compounds to carbon dioxide and water (Fig. 2).

During strenuous exercise breathing and heart rate increase, but the volume of oxygen which they deliver to the muscles is still very small compared with the volume actually required to completely break down food to carbon dioxide and water. Under these conditions energy for muscular contraction is obtained mainly from the anaerobic stage of respiration. This produces lactic acid faster than it can be broken down by the aerobic stage of respiration, and so lactic acid accumulates in the muscles.

The body of a trained athlete can tolerate about 127 g of lactic acid before the acid prevents further muscular effort, and for every 10 g of lactic acid in the body 1·7 litres of oxygen must be absorbed through the lungs to remove the acid. This oxygen requirement is called the **oxygen debt**. The 'debt' is 'paid' by rapid breathing for a time after the exercise stops (Fig. 3). In other words, during strenuous exercise the body produces energy first and obtains oxygen to 'pay' for it later.

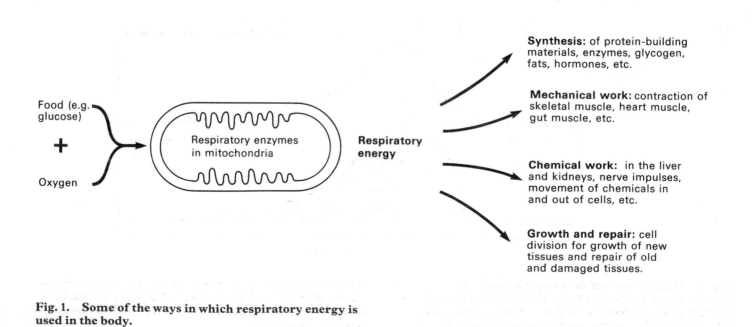

Fig. 1.   Some of the ways in which respiratory energy is used in the body.

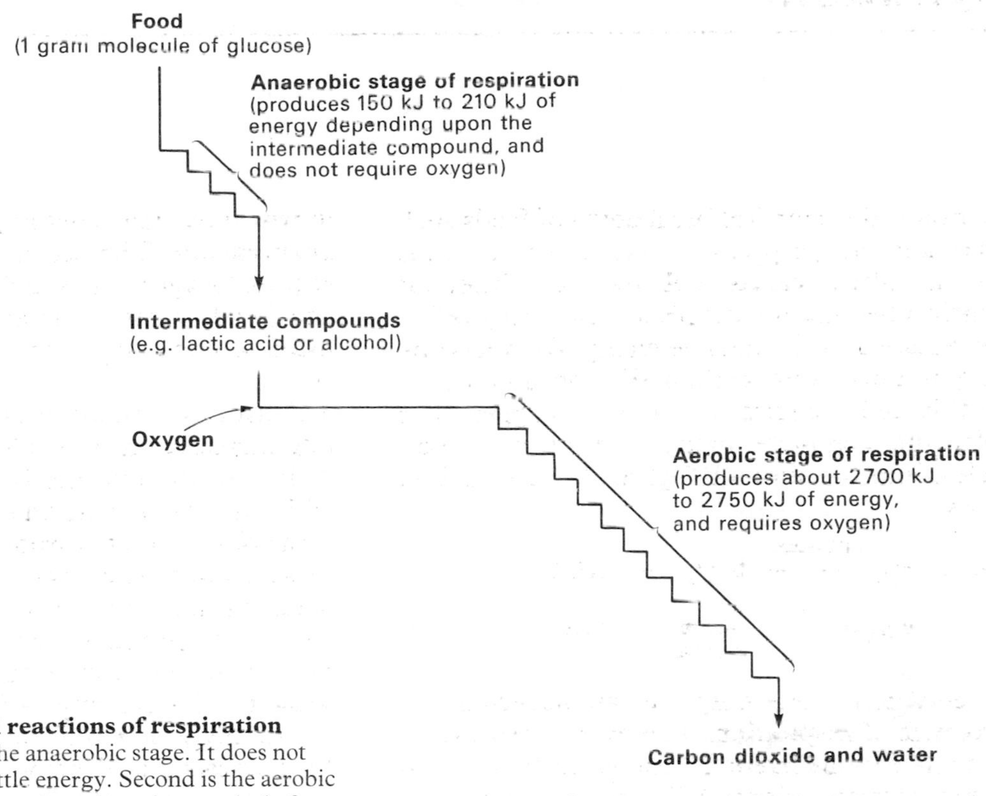

Food
(1 gram molecule of glucose)

Anaerobic stage of respiration
(produces 150 kJ to 210 kJ of
energy depending upon the
intermediate compound, and
does not require oxygen)

Intermediate compounds
(e.g. lactic acid or alcohol)

Oxygen

Aerobic stage of respiration
(produces about 2700 kJ
to 2750 kJ of energy,
and requires oxygen)

Carbon dioxide and water

Fig. 2. The many chemical reactions of respiration occur in two stages. First is the anaerobic stage. It does not require oxygen and produces little energy. Second is the aerobic stage. This does require oxygen and produces a great deal of energy.

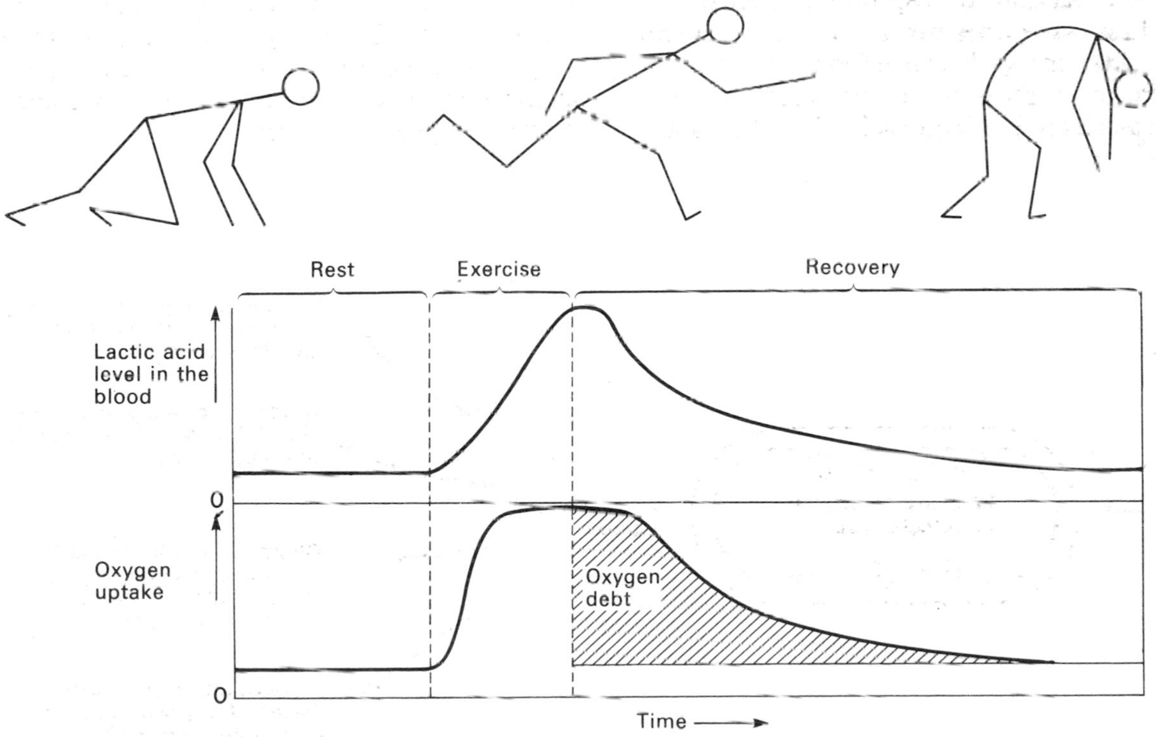

Rest   Exercise   Recovery

Lactic acid level in the blood

0

Oxygen uptake

Oxygen debt

0

Time ⟶

Fig. 3. Respiration and the oxygen debt. During strenuous exercise oxygen uptake rises to a peak (bottom graph) but this is not enough to completely break down food in the muscles to CO_2 and H_2O. So anaerobic respiration increases, leading to a build-up of lactic acid (top graph). When exercise stops oxygen uptake continues at a high level until all the lactic acid has been oxidized. Thus the oxygen 'debt' is paid by rapid breathing when the exercise stops

6

Adenosine triphosphate (ATP)

The energy released from food during respiration cannot be used directly by the body. The energy is used to create molecules of a chemical called **adenosine triphosphate**, or **ATP**. ATP is a temporary store of energy, which can be released whenever required for a wide variety of jobs, such as contraction of muscle or synthesis of complex chemicals.

ATP is formed during respiration from a related substance called **adenosine diphosphate**, or **ADP**. ADP consists of adenosine plus two phosphate groups bonded to it (Fig. 2A). The energy released during respiration is used to bond a third phosphate group to ADP molecules making ATP. It requires a considerable amount of energy to do this, and so the process is known as the formation of **energy-rich phosphate bonds** (Fig. 2 B and C). These energy-rich bonds are easily broken, and this happens whenever energy is required. When a bond is broken the same amount of energy is released as went into its creation.

As ATP parts with its third phosphate group it becomes ADP once more. This ADP is used to make more ATP and the process repeats itself over and over again.

There are four main advantages to the ADP/ATP system. First, ATP takes up some of the energy which would otherwise be lost as heat during the breakdown of food by respiratory enzymes. This increases the efficiency of respiration. Second, energy can be released from ATP the instant it is required without having to go through the fifty different reactions of respiration. This is very important in an emergency, or during strenuous exercise, when a sudden burst of energy is required. Figure 3 describes an experiment to show that ATP releases energy which makes muscles contract. Third, ATP delivers energy in precisely controlled amounts, since each energy-rich phosphate bond releases a specific amount of energy when broken (about 34 kJ per mole of ATP). Fourth, energy-rich bonds can be transferred from ATP to other substances without any loss of energy, converting these substances from a relatively-inert state to a highly reactive one. This is very important in the synthesis of complex chemicals from simpler ones.

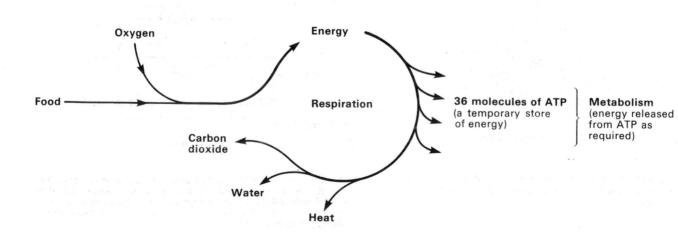

Fig. 1. Energy released during respiration is not used directly, it is used to create molecules of adenosine triphosphate, or ATP. ATP stores the energy temporarily until it is required for metabolism.

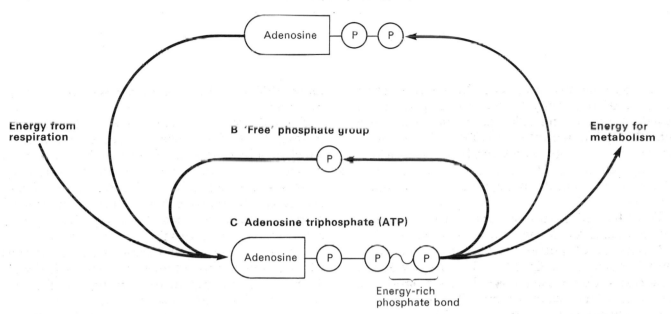

A Adenosine diphosphate (ADP)

B 'Free' phosphate group

C Adenosine triphosphate (ATP)

Energy from respiration

Energy for metabolism

Energy-rich phosphate bond

Fig. 2. ATP is made from ADP. ADP consists of adenosine and two phosphate groups. Energy from respiration links a third phosphate onto ADP by forming an energy-rich bond, making ATP. When this bond is broken the same amount of energy is released as went into its construction (about 34 kJ per mole of ATP). Then ATP reverts to ADP and 'free' phosphate which are later used to make more ATP.

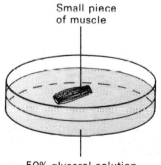

Small piece of muscle

50% glycerol solution

A Cut a small piece of muscle from a freshly killed animal and soak it in 50% glycerol solution for two days in a refrigerator

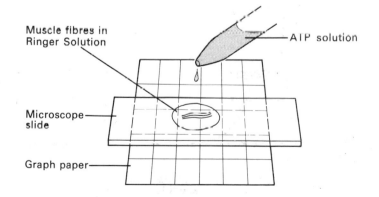

Muscle fibres in Ringer Solution

ATP solution

Microscope slide

Graph paper

B Wash the muscle in Ringer Solution. Take a few fibres from the muscle and spread them out straight on a microscope slide in a little Ringer Solution. Put the slide on graph paper and measure the fibres. Drop ATP solution onto the fibres and observe any change in their length

Fig. 3. Muscle fibres contract in ATP solution.

Epithelial tissues

What are tissues?

Animals are multicellular. This means they consist of many cells. But their cells are not all alike. Most cells are specialized to perform a certain function. Nerve cells, for example, are specialized to conduct nerve impulses, and muscle cells are specialized for contraction. Groups of specialized cells are called **tissues**. There are four types of tissue: epithelial, muscular, nervous, and connective.

Epithelial tissues

These tissues consist of sheets of cells which cover the external and internal surfaces of the body, and line the insides of glands.

The blood vessels and heart are lined with a single layer of flattened cells called **squamous**, or **pavement**, **epithelium**, because the cells are shaped like paving stones (Fig. 1). This tissue also covers the heart, lungs, and other organs and lines the spaces in which the organs lie, reducing friction between surfaces which rub together.

The ducts (small tubes) in the kidneys and those leading from the salivary glands are lined with a single layer of box-shaped cells called **cubical epi-**

thelium (Fig. 2B). The large kidney ducts, the gall bladder, stomach, and small intestine are lined with a single layer of elongated cells called **columnar epithelium** (Fig. 2E). The passageways of the respiratory and reproductive systems are lined with both cubical and columnar cells covered with microscopic hairs called **cilia** (Fig. 2A). The cilia of this **ciliated epithelium** wave rapidly back and forth creating currents in liquids surrounding them.

The glands of the body are lined with cubical or columnar cells called **glandular epithelium**. These cells are specialized to produce useful substances like enzymes, mucus, and hormones (Fig. 2 B and E).

The outer surface of the body, and the mouth and the gullet, are exposed to constant wear and tear. These surfaces are protected from damage and infection by several layers of cells called **stratified epithelium** (Fig. 2D). The uppermost layer of stratified epithelium consists of flat dead cells which are constantly worn away. But as fast as this happens they are replaced from below by new cells from a layer of live dividing cells.

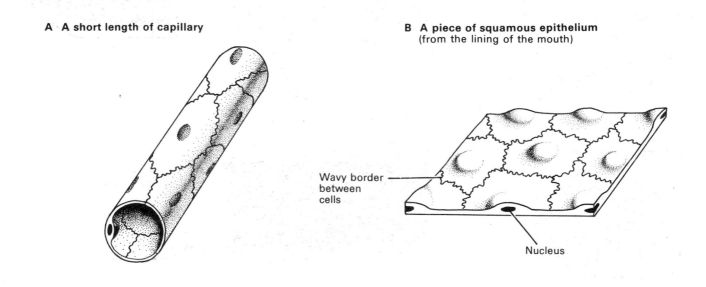

A A short length of capillary

B A piece of squamous epithelium
(from the lining of the mouth)

Wavy border between cells

Nucleus

Fig. 1. Squamous, or pavement, epithelium, consists of a single layer of flat cells. It forms the walls of capillaries, lines the mouth and heart, and covers body organs.

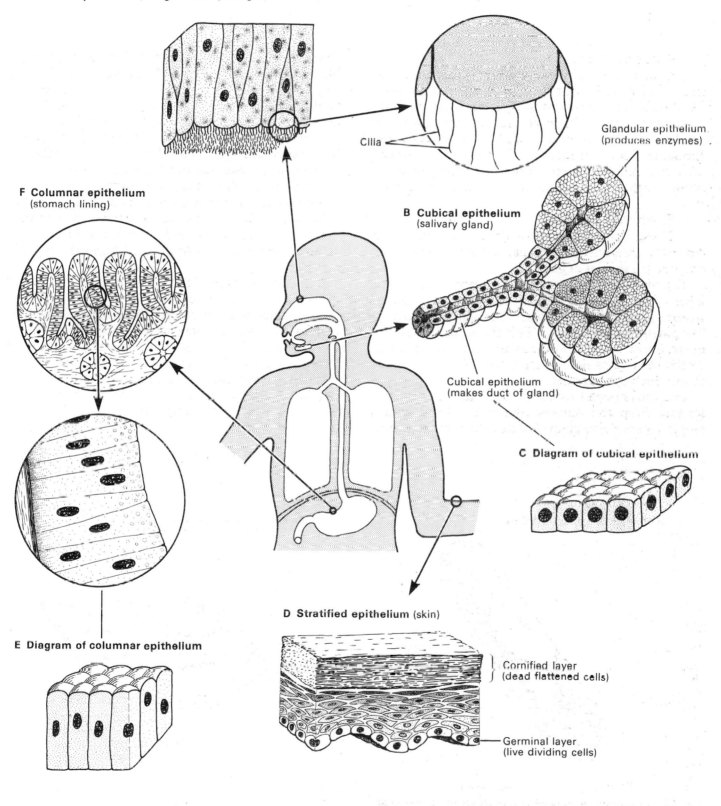

A Ciliated epithelium (lining of nasal passages)

Cilia

Glandular epithelium (produces enzymes)

F Columnar epithelium (stomach lining)

B Cubical epithelium (salivary gland)

Cubical epithelium (makes duct of gland)

C Diagram of cubical epithelium

E Diagram of columnar epithelium

D Stratified epithelium (skin)

Cornified layer (dead flattened cells)

Germinal layer (live dividing cells)

Fig. 2. More examples of epithelial tissue. The tissues shown in **A, B, C, E,** and **F** consist of a single layer of either cubical or oblong (columnar) cells which cover internal body surfaces. Tissue **D** covers the outside of the body.

8
Muscle tissues

Muscles are specialized for contraction. So muscle tissue consists of contractile fibres. These fibres are bound together by connective tissues which also connect muscles with bones or skin. Contraction of a muscle fibre is caused by a complex rearrangement of its protein molecules. The energy needed for this contraction comes from respiration.

An extensive network of blood capillaries runs between the muscle fibres, and there are also two types of nerve fibre. Motor nerves cause muscles to contract, and sensory nerves detect changes in the length of muscle fibres. These types of nerve are described in Unit 9.

There are three types of muscle tissue, and each performs a different type of contraction.

1. Muscle found in the walls of the gut, blood vessels, and bladder is called **involuntary muscle tissue** because it cannot be contracted at will. Involuntary muscles are controlled by unconscious mechanisms described in Unit 53. Involuntary muscle fibres are single cells which are elongated and pointed at each end (Fig. 1). The fibres contract slowly and rhythmically, and can remain contracted for long periods. They control processes like peristalsis which moves food along the gut.

2. **Cardiac muscle tissue** is found only in the heart. It consists of single cells which are branched and joined together making a network of fibres (Fig. 2). These fibres contract rhythmically with moderate speed throughout life, pumping blood around the body.

3. The type of muscle which is attached to the skeleton is called **voluntary muscle tissue** because it can be contracted at will (voluntarily). It is also called striped muscle because of its striped appearance (Fig. 3B). Voluntary muscle fibres are up to 40 mm in length. They consist of many cells and their structure is very complicated. Voluntary muscle can contract quickly and powerfully and causes movements of the body.

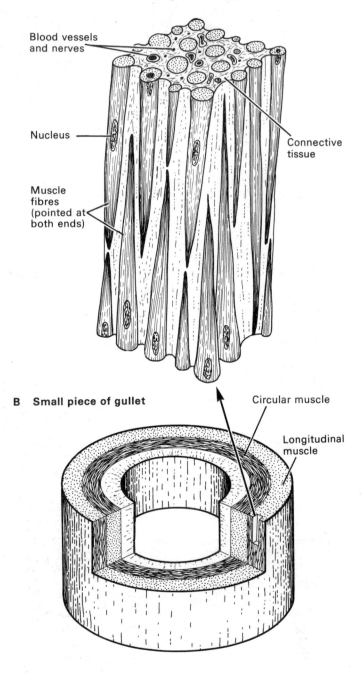

A Involuntary muscle

Blood vessels and nerves

Nucleus

Connective tissue

Muscle fibres (pointed at both ends)

B   Small piece of gullet

Circular muscle

Longitudinal muscle

Fig. 1. Involuntary muscle tissue is found in the wall of the gullet (oesophagus). The muscle fibres are single cells which contract slowly and rhythmically, moving food along the gullet.

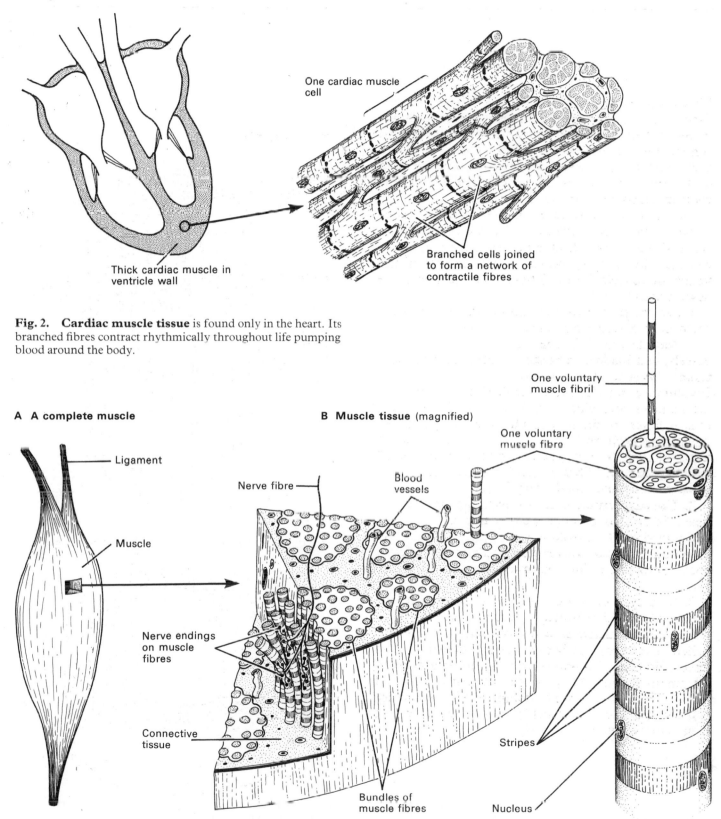

A Diagram of the heart

B Cardiac muscle fibres

One cardiac muscle cell

Thick cardiac muscle in ventricle wall

Branched cells joined to form a network of contractile fibres

Fig. 2. **Cardiac muscle tissue** is found only in the heart. Its branched fibres contract rhythmically throughout life pumping blood around the body.

A A complete muscle

B Muscle tissue (magnified)

One voluntary muscle fibril

One voluntary muscle fibre

Ligament

Nerve fibre

Blood vessels

Muscle

Nerve endings on muscle fibres

Connective tissue

Bundles of muscle fibres

Stripes

Nucleus

Fig. 3. **Voluntary, or skeletal, muscle tissue** is attached to bones. Its long striped fibres contract quickly and powerfully causing movements of the body.

17

9

Nervous tissues and nerve impulses

Nervous tissues contain cells which conduct 'messages' called **nerve impulses**. These cells are called **neurones**. The spaces between neurones are filled with cells called **neuroglia**, which support neurones, nourish them, and help them conduct impulses.

Neurones and nerve fibres

Like other cells neurones consist of a nucleus and cytoplasm. But unlike other cells the cytoplasm of a neurone extends outwards forming fine threads called **nerve fibres**. A nerve is a bundle of nerve fibres bound together by connective tissue (Fig. 3A). Nerve fibres can be over one metre in length, and between 0·004 mm and 0·02 mm thick. The thicker the fibre the faster impulses travel along it. The thickest fibres conduct impulses at 120 metres per second. Nerve fibres outside the brain and spinal cord are enclosed in a sheath of fatty insulating material called **myelin**. This material is contained in the cytoplasm of special cells which wrap themselves around the nerve fibres (Fig. 2C).

Nerve cells which conduct impulses from the sense organs (eyes, ears, etc.) to the brain and spinal cord are called **sensory neurones** (Figs. 1 and 2A). Inside the brain and spinal cord impulses pass from sensory neurones to **association neurones** which, in turn, pass them to **motor neurones** (Fig. 1). Motor neurones (Fig. 2B) conduct impulses out of the brain and spinal cord to the muscles and glands.

Nerve impulses

Light, sound, touch, and heat are called **stimuli** (*singular* stimulus) because they stimulate sense organs into sending impulses along nerves. An impulse can be described as a travelling wave of electrical and chemical changes. There is no such thing as a strong or a weak impulse, and there is no difference between impulses from eyes, ears, or other sense organs. All impulses are exactly alike. The only feature of impulses which varies is the number per second travelling along a fibre. A very strong stimulus, such as a loud noise, can cause up to 100 impulses per second to travel from the ears to the brain. Alternatively there are sounds too faint to stimulate the ears. Stimulation must reach what is called the **threshold level** of a sense organ before it begins sending out impulses.

Each neurone is connected with many other neurones making literally millions of interconnections within the nervous system. But neurones are not continuous with one another. There is a gap called a **synapse** where they meet (Fig. 3 B and C). Impulses arriving at a synapse cause the release of chemicals which diffuse across the gap and stimulate the production of impulses in the next neurone.

Each synapse is a barrier to nerve impulses, and cannot be crossed unless impulses are arriving there at a certain number per second. These millions of barriers ensure that weak stimuli are ignored, and prevent impulses in one fibre from running through nerve interconnections to stimulate all the neurones in the body.

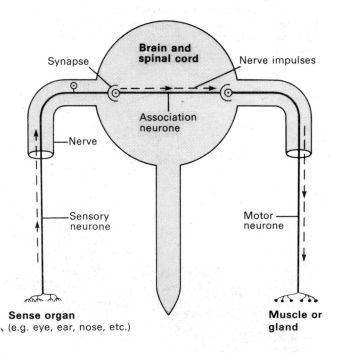

Fig. 1. Diagram of how nerve cells (neurones) are arranged in the nervous system. Sensory neurones carry impulses from sense organs to the brain or spinal cord. Association neurones carry impulses through the brain or spinal cord and pass them to motor neurones which carry them to muscles or glands.

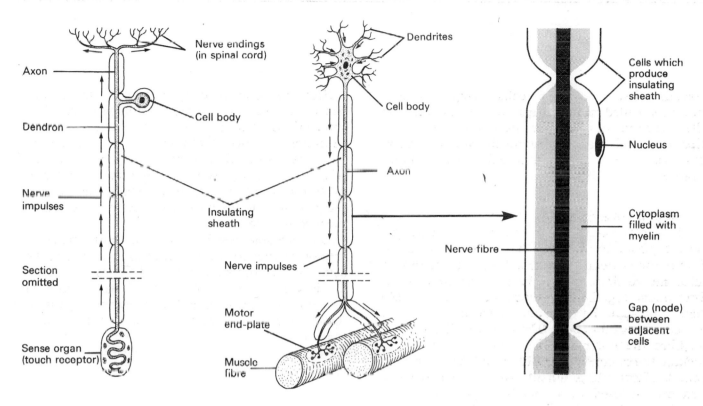

A A typical sensory neurone

Nerve endings
(in spinal cord)

Axon

Cell body

Dendron

Nerve
impulses

Insulating
sheath

Section
omitted

Sense organ
(touch receptor)

B A typical motor neurone

Dendrites

Cell body

Axon

Nerve impulses

Motor
end-plate

Muscle
fibre

C Part of a nerve fibre (magnified)

Cells which
produce
insulating
sheath

Nucleus

Cytoplasm
filled with
myelin

Nerve fibre

Gap (node)
between
adjacent
cells

Fig. 2 A and B.   Sensory and motor neurones. Neurones consist of a cell body and nerve fibres. Depending upon its length, a fibre which carries impulses towards the cell body is called a dendron (long) or dendrite (short). An axon carries impulses away from the cell body

Fig. 2C.   Nerve fibres outside the brain and spinal cord are covered by a sheath of fatty insulating material called myelin. The myelin is produced by cells which wrap themselves around the nerve fibres.

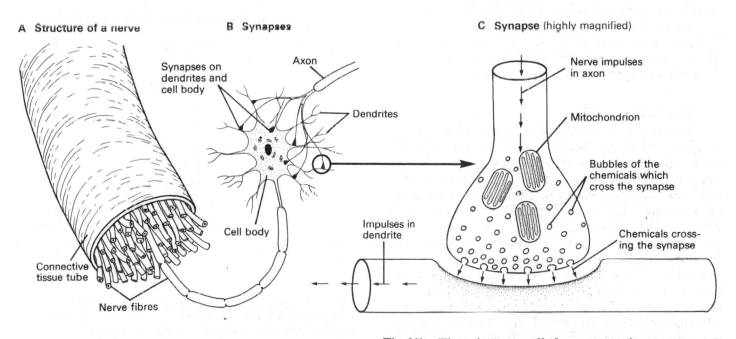

A Structure of a nerve

Connective
tissue tube

Nerve fibres

B Synapses

Synapses on
dendrites and
cell body

Axon

Dendrites

Cell body

Impulses in
dendrite

C Synapse (highly magnified)

Nerve impulses
in axon

Mitochondrion

Bubbles of the
chemicals which
cross the synapse

Chemicals cross-
ing the synapse

Fig. 3 A and B.   A nerve is a bundle of nerve fibres in a tube of connective tissue.

Fig. 3C.   There is a gap called a synapse where one nerve fibre meets another. Impulses arriving at a synapse release chemicals which cross the gap and stimulate the next neurone.

19

Connective tissues

All connective tissues consist of cells embedded in an intracellular substance called the **matrix**, which the cells produce. As their name suggests these tissues connect tissues and organs together, protecting and supporting them, and yet allowing them to move against one another. Some connective tissues are hard and give more rigid support. There are four types of connective tissue.

1. The spaces between organs are packed with **areolar connective tissue**. This consists of fibres and cells enclosed in a white sticky matrix (Fig. 1). The matrix is produced by **mast cells**, and these also produce heparin, a chemical which helps stop blood from clotting inside blood vessels. **Macrophage cells** engulf and digest bacteria which enter wounds, and they can move through the matrix to infected areas. **Fibroblast cells** make the fibres of areolar and other tissues. They too can move to infected areas and seal off wounds or enclose parasites in a mass of fibres.

2. Tendons and ligaments are made of **fibrous connective tissue**. Tendons connect muscles to bones and are made of tough, white **collagen fibres** (Fig. 4). Ligaments hold joints together, and are made of **elastic fibres**. These fibres are also found in blood vessel walls (Fig. 2).

3. The gristle which occurs at the ends of some bones, where bones rub together, and in the walls of the wind-pipe is made of **cartilage** (Fig. 3). Cartilage consists of cells called **chondroblasts**, which are embedded in a clear bluish coloured matrix that is hard and yet slightly flexible. This tissue is often filled with collagen and elastic fibres, when it is called **fibro-cartilage**.

4. Bone is a connective tissue which consists of cells in a solid matrix impregnated with calcium, phosphate, carbonate, and fluoride salts. Bone cells are usually arranged in concentric circles around blood vessels which run in narrow channels through the length of a bone (Fig. 4). A system of tiny channels connects the blood canals with the cavities in which the bone cells live. These channels deliver food and oxygen from the bloodstream to the bone cells and carry away waste matter.

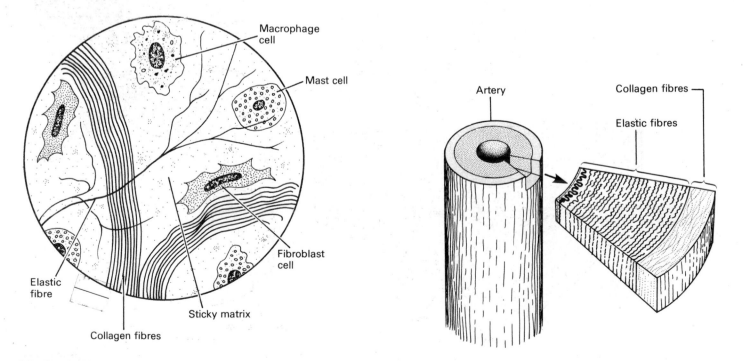

Fig. 1. Areolar connective tissue (seen through a microscope).

Fig. 2. Elastic and collagen fibres in a blood vessel wall.

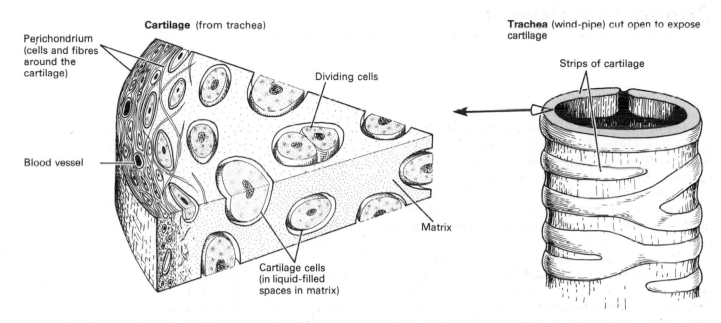

Cartilage (from trachea)

Perichondrium (cells and fibres around the cartilage)

Dividing cells

Blood vessel

Matrix

Cartilage cells (in liquid-filled spaces in matrix)

Trachea (wind-pipe) cut open to expose cartilage

Strips of cartilage

Fig. 3.   Structure of cartilage.

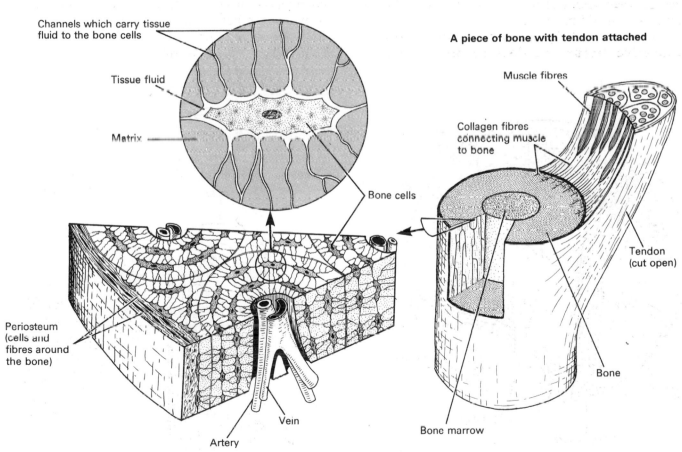

A bone cell (highly magnified)

Channels which carry tissue fluid to the bone cells

Tissue fluid

Matrix

Bone cells

A piece of bone with tendon attached

Muscle fibres

Collagen fibres connecting muscle to bone

Tendon (cut open)

Periosteum (cells and fibres around the bone)

Vein

Artery

Bone

Bone marrow

Fig. 4.   Structure of a bone.

11

Organs and organ systems

An **organ** consists of several different tissues grouped together. The heart, for example, is made up of cardiac muscle and nervous tissues bound together with connective tissues, and it is lined with squamous epithelium.

Several organs working together form an **organ system**. The urinary system, for example, consists of the kidneys, bladder, and ureters. The Figures of this unit illustrate all the main organ systems of the human body.

Most organs are situated in the **body cavities**. The main body cavities are the **thorax**, which is the chest cavity above the diaphragm (Fig. 3), and the **abdomen**, which is below the diaphragm.

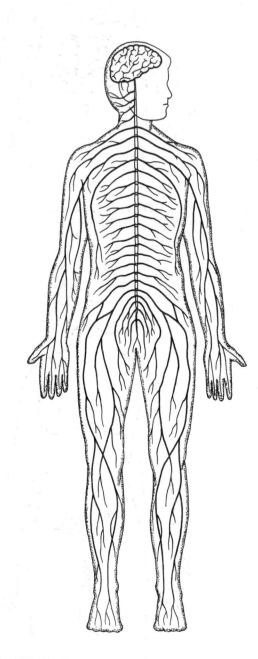

Fig. 1. Nervous system.

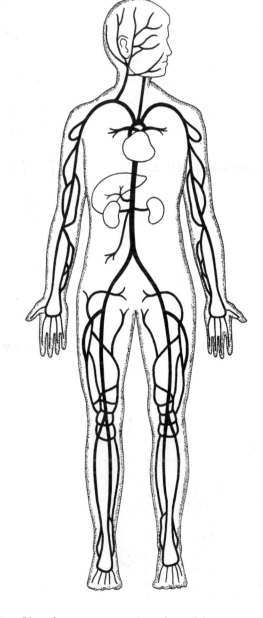

Fig. 2. Circulatory system (arteries only).

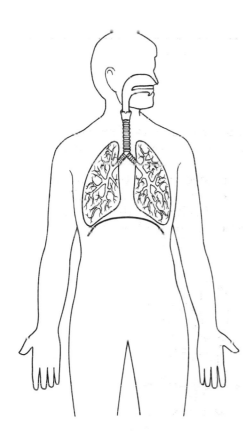

Fig. 3. Respiratory system.

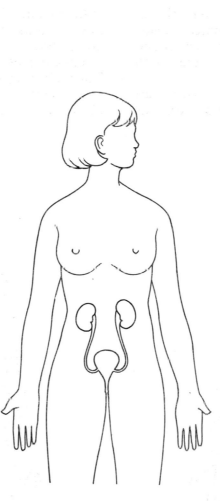

Fig. 5. Female urinary system.

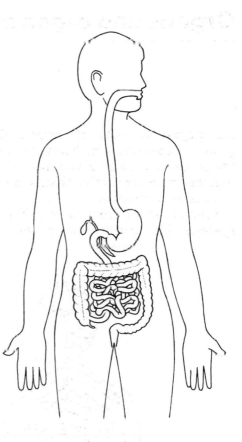

Fig. 4. Digestive system.

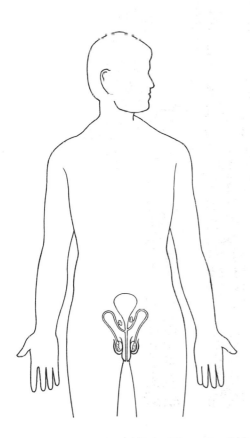

Fig. 6. Male reproductive system.

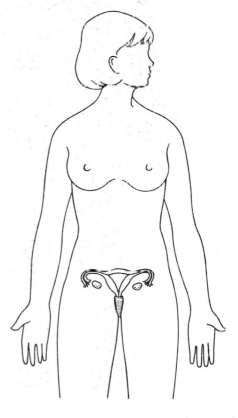

Fig. 7. Female reproductive system.

The skeleton and its functions

Humans and all other vertebrates have skeletons based on the following plan. The skull, backbone or **vertebral column**, and rib cage form the main axis of the body and so are called the **axial skeleton**. The pectoral and pelvic girdles (shoulder blades and hips) and the arm and leg bones are attached (appended) to the axial bones and so are called the **appendicular skeleton** (Fig. 1).

The backbone consists of small bones called **vertebrae**. There are thirty-three vertebrae in an embryo but some fuse together shortly after birth leaving twenty-six in adults. Each vertebra has a large hole called the **neural canal** which contains and protects the spinal cord (Figs. 3 and 4). Vertebrae also have a number of **processes** to which muscles are attached. Special processes with flat surfaces form the joints where adjacent vertebrae rub together. The shape and functions of the vertebrae vary depending upon their position in the backbone (Fig. 2). Vertebrae are separated by discs of cartilage which act as shock absorbers preventing shock waves from jarring the backbone during running and jumping (Fig. 4).

Bones form a rigid framework which supports soft tissues against the force of gravity and maintains the shape of the body. Most organs are suspended from the skeleton and this allows them to function freely without becoming entangled or crushing each other.

The brain box or **cranium** of the skull encloses and protects the brain, the inner and middle ears, and nasal organs. The eyes are partly enclosed in sockets of bone called **orbits** (Fig. 1). The heart, lungs, and major blood vessels are protected by the rib cage.

Bones and joints form a system of levers which are moved by the muscles. Muscles contract and pull against the bones causing movement at the joints. This allows the whole body or its parts to be moved at will.

Muscles between the ribs raise and lower the rib cage during deep breathing. In addition, the rib cage supports the diaphragm muscle which is also concerned with breathing.

Red and white blood cells and platelets are manufactured in the **bone marrow**. Marrow tissue is concentrated inside the long limb bones and ribs.

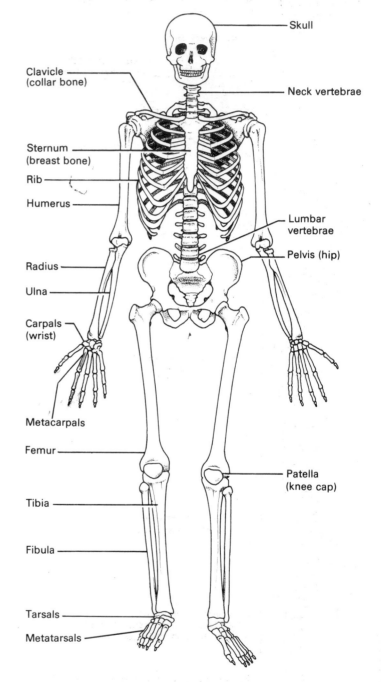

Fig. 1.   The skeleton protects and supports the soft parts of the body. The skull protects the brain, inner and middle ears, and nasal organs. Eye sockets protect the eyes. The backbone protects the nerve cord; and the heart, lungs, and main blood vessels are protected by the rib cage. Bones, joints, and muscles make movement possible. Blood cells are made in the marrow at the centre of the long bones.

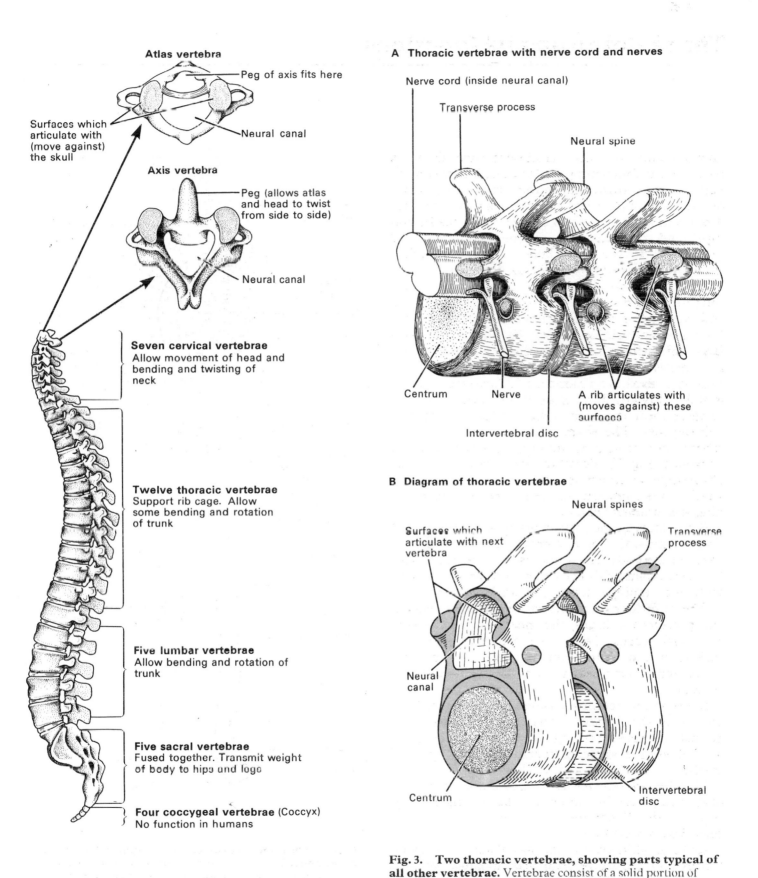

Atlas vertebra

Peg of axis fits here

Surfaces which articulate with (move against) the skull

Neural canal

Axis vertebra

Peg (allows atlas and head to twist from side to side)

Neural canal

Seven cervical vertebrae
Allow movement of head and bending and twisting of neck

Twelve thoracic vertebrae
Support rib cage. Allow some bending and rotation of trunk

Five lumbar vertebrae
Allow bending and rotation of trunk

Five sacral vertebrae
Fused together. Transmit weight of body to hips and legs

Four coccygeal vertebrae (Coccyx)
No function in humans

A Thoracic vertebrae with nerve cord and nerves

Nerve cord (inside neural canal)

Transverse process

Neural spine

Centrum

Nerve

A rib articulates with (moves against) these surfaces

Intervertebral disc

B Diagram of thoracic vertebrae

Neural spines

Surfaces which articulate with next vertebra

Transverse process

Neural canal

Centrum

Intervertebral disc

Fig. 2. Side view of the human backbone (vertebral column). It consists of thirty-three bones, some of which are fused together. The shape and functions of vertebrae vary depending upon their position in the backbone.

Fig. 3. Two thoracic vertebrae, showing parts typical of all other vertebrae. Vertebrae consist of a solid portion of bone called the centrum, above which is the canal that contains and protects the spinal cord. There are a number of processes to which muscles are attached, and four processes with flat surfaces which form joints where adjacent vertebrae rub together.

25

13

Structure and growth of bones

Bones are encased in a layer of tough fibres and cells called the **periosteum** (Fig. 1). Underneath this is a layer of hard **compact bone**, whose structure is described in Unit 10. Compact bone gives the skeleton its strength. Next is a layer of **spongy bone**, so called because of its many tiny cavities and not because it is soft. At the centre of most bones is a **marrow cavity**. This cavity is filled with fat, and with the tissues which make red and white blood cells.

Bones such as the thigh bone (femur) consist of a long hollow shaft or **diaphysis**, and two rounded ends or 'heads' called **epiphyses** (Fig. 1). The heads are mostly spongy bone covered with slippery cartilage where they rub against another bone at a joint. The shaft is a cylinder of compact bone with spongy bone and marrow at its centre.

Most of the skeleton of a developing embryo is formed of cartilage. During subsequent development this cartilage skeleton is converted into bone by a process called **ossification**. The ossification of a femur, for example, occurs in the following way.

1. Bone-forming cells develop around the middle of the femur's shaft and form a thin cylinder of bone which supports the remaining cartilage as it is replaced by bone (Fig. 2 A and B).

2. Cartilage in the centre of the shaft and heads becomes saturated with calcium salts, a process called **calcification**. Cartilage cells then swell up and die leaving rows of empty holes (Fig. 2B).

3. Blood vessels and bone-forming cells from the surface of the femur grow into the bone (Fig. 2C). They open up the rows of holes left by cartilage cells, forming tunnels lengthwise along the bone (Fig. 2D). Bone-forming cells lay down concentric rings of bone around the blood vessels forming bone tissue, described in Unit 10.

4. A strip of cartilage remains between the shaft and the head of the femur (Figs. 1 and 2D). Cell division in this strip adds new cartilage to the shaft, increasing the length of the bone as the child grows. This new cartilage is ossified as fast as it forms, and when growth stops the cartilage strip itself is ossified. A bone is increased in width by new bone formed under its periosteum. Throughout life bones are constantly remoulded by cells which destroy bone and others which rebuild it, so that their shape is perfectly suited to their function.

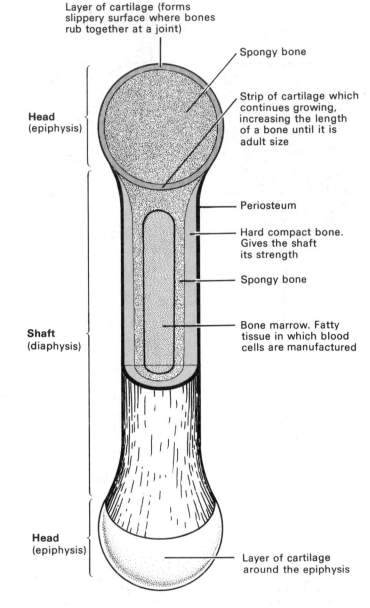

Layer of cartilage (forms slippery surface where bones rub together at a joint)

Spongy bone

Strip of cartilage which continues growing, increasing the length of a bone until it is adult size

Head (epiphysis)

Periosteum

Hard compact bone. Gives the shaft its strength

Spongy bone

Shaft (diaphysis)

Bone marrow. Fatty tissue in which blood cells are manufactured

Head (epiphysis)

Layer of cartilage around the epiphysis

Fig. 1. Diagram illustrating the structure of a long bone, such as the femur (the top half is cut open lengthwise). These bones consist of a long hollow shaft and two rounded heads. The heads are covered with slippery cartilage where they rub against another bone at a joint. The heads contain 'spongy' bone full of tiny cavities. The shaft is a hollow cylinder of hard compact bone filled with spongy bone and marrow tissue at its centre.

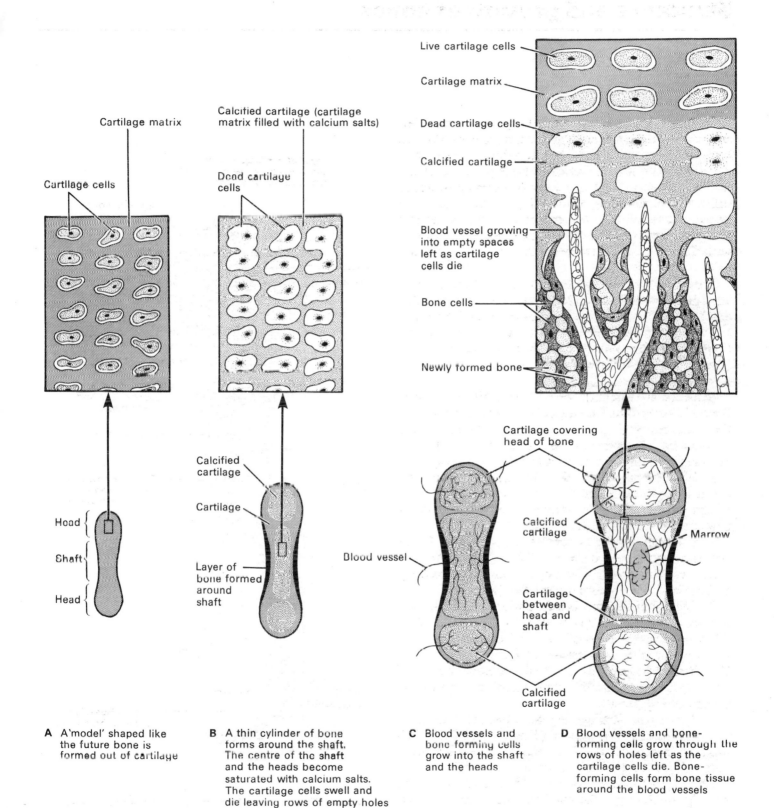

Cartilage matrix

Cartilage cells

Calcified cartilage (cartilage matrix filled with calcium salts)

Dead cartilage cells

Head {
Shaft {
Head {

Calcified cartilage

Cartilage

Layer of bone formed around shaft

Live cartilage cells

Cartilage matrix

Dead cartilage cells

Calcified cartilage

Blood vessel growing into empty spaces left as cartilage cells die

Bone cells

Newly formed bone

Cartilage covering head of bone

Blood vessel

Calcified cartilage

Marrow

Cartilage between head and shaft

Calcified cartilage

A A 'model' shaped like the future bone is formed out of cartilage

B A thin cylinder of bone forms around the shaft. The centre of the shaft and the heads become saturated with calcium salts. The cartilage cells swell and die leaving rows of empty holes

C Blood vessels and bone forming cells grow into the shaft and the heads

D Blood vessels and bone-forming cells grow through the rows of holes left as the cartilage cells die. Bone-forming cells form bone tissue around the blood vessels

Fig. 2. The early development of a long bone (such as a femur). At first they are made of cartilage. The process by which this cartilage is converted into bone is called ossification. Drawings **A–D** above illustrate the stages of ossification.

14

Joints

Joints occur wherever two or more bones touch. In some joints the bones are bound firmly together by tough fibres so that no movement is possible. These are called **fixed joints**. Examples occur between the bones which form the roof of the skull (Fig. 12.1). In **slightly movable** joints the bones can move a little against a pad of cartilage situated between them. Vertebrae, for example, can move slightly against the discs of cartilage inbetween them (Fig. 12.3).

There are about seventy **freely movable**, or **synovial joints** in the skeleton. The shoulder, knee, and elbow are examples. Where the bones of a freely movable joint rub together they are covered with shiny, slippery cartilage. These joints are lubricated by a liquid called **synovial fluid**. This fluid is sealed in by the **synovial membrane**, which surrounds the whole joint (Fig. 1). The bones of freely movable joints are held in place and yet allowed to move freely by bands of fibre called **ligaments** (Fig. 2).

The type of movement possible at a freely movable joint depends upon the shape of the bones at the point where they rub together.

In **hinge joints**, such as the elbows and knees, movement can occur in only one direction, like the hinge of a door (Fig. 3B).

In **ball-and-socket** joints such as the hips and shoulders, the rounded head of one bone fits into a cup-shaped socket of another. These joints allow movement in many directions (Fig. 3C).

In **pivot joints** one bone twists against another. A pivot joint occurs where a peg on the axis vertebra fits into a socket in the atlas vertebra (Fig. 3D).

In a **sliding** or **gliding joint** the surfaces which rub together are flat. Examples occur between the vertebrae where two flat-surfaced projections from each vertebra move against each other when the backbone twists or bends (Fig. 3A).

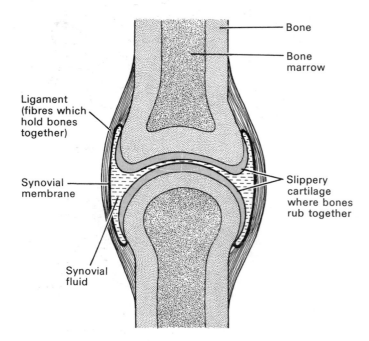

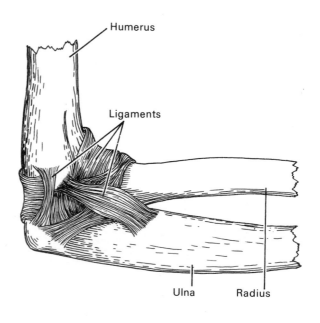

Fig. 1. Freely movable, or synovial, joints have a structure which reduces friction where the bones rub together. Here, the bones are covered with shiny, slippery cartilage and the joint is lubricated with oily synovial fluid.

Fig. 2. The ligaments which hold the elbow joint together. Ligaments are tough fibres which hold bones in place at a joint and yet allow the joint to move freely.

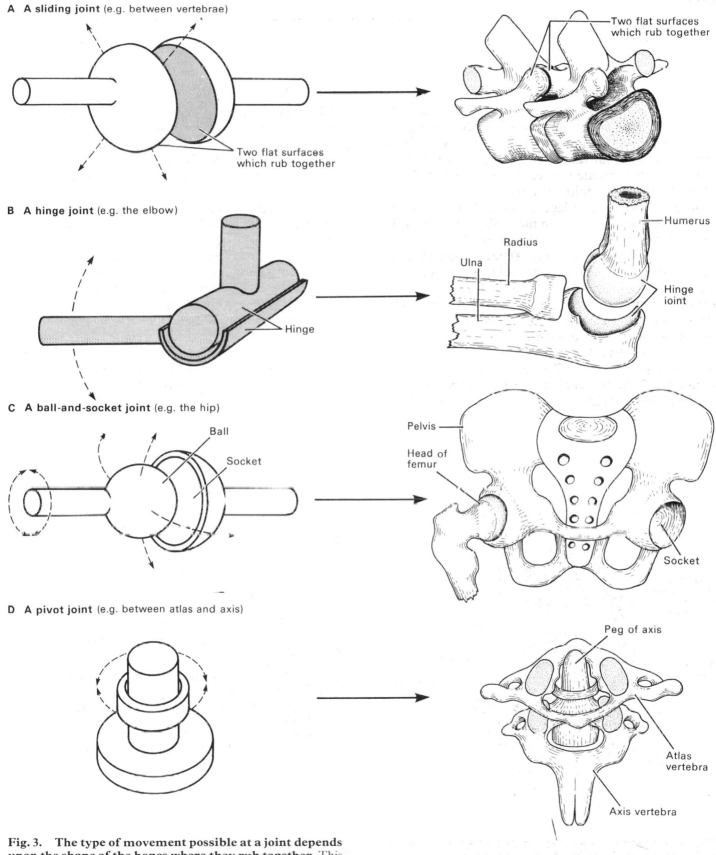

A A sliding joint (e.g. between vertebrae)

Two flat surfaces which rub together

Two flat surfaces which rub together

B A hinge joint (e.g. the elbow)

Hinge

Radius

Ulna

Humerus

Hinge joint

C A ball-and-socket joint (e.g. the hip)

Ball

Socket

Pelvis

Head of femur

Socket

D A pivot joint (e.g. between atlas and axis)

Peg of axis

Atlas vertebra

Axis vertebra

Fig. 3. The type of movement possible at a joint depends upon the shape of the bones where they rub together. This diagram illustrates some examples. Experiment carefully with these joints to discover the ways in which they can move.

Muscles, movement, and leverage

The microscopic structure of muscle is described in Unit 8. This Unit describes muscles in action.

Muscles cause movement of the body by contracting. That is, they become shorter in length. As this happens, they pull against the skeleton making it bend or straighten at a joint (Fig. 1).

Muscles are attached to bones by **tendons**, described in Unit 10. The tendon at the end of a muscle closest to the joint it moves is called the **insertion** of that muscle (Fig. 2). The insertion moves during muscular contraction while at the opposite end, called the **origin** of the muscle, is the tendon which remains fixed in position. This tendon is the anchorage point of the muscle. The biceps muscle of the arm, for example, has two origins on the shoulder blade, and one insertion on the radius bone.

Muscles can pull but they cannot push. Therefore a muscle which moves a bone in one direction cannot move it back the other way. Another muscle has to pull it in the opposite direction. Consequently, there must be at least two muscles at each joint. One, called the **flexor muscle**, contracts and bends a joint (Fig. 1A). On the opposite side of the joint there is an **extensor muscle**. This straightens the joint (Fig. 1B). Since flexor and extensor muscles pull a joint in opposite directions they are said to form an **antagonistic system** (Figs. 1 and 2).

During movements bones act as levers. The point where a lever moves or pivots is called the **fulcrum**. In a skeleton, a joint acts as a fulcrum. The force which moves a lever is known as **effort**. In a skeleton, effort is supplied by muscles. Levers can carry or move a **load** of some kind. The load which the skeleton carries or moves is the weight of the body, and any additional weight supported by the body.

There are three types of lever system. In each type the fulcrum, effort, and load are in a different position. Figure 3 illustrates an example of each type of lever from the human skeleton.

A Action of a flexor muscle

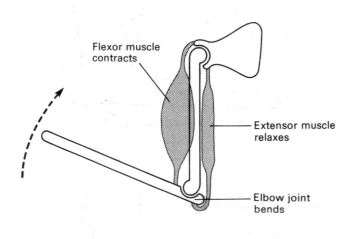

Flexor muscle contracts

Extensor muscle relaxes

Elbow joint bends

B Action of an extensor muscle

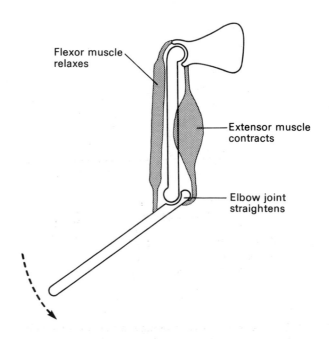

Flexor muscle relaxes

Extensor muscle contracts

Elbow joint straightens

Fig. 1. Muscles pull against the skeleton, bending or straightening it at the joints. Muscles which bend joints are called flexor muscles, and muscles which straighten joints are called extensor muscles. Since flexor and extensor muscles pull in opposite directions they are said to form an antagonistic system.

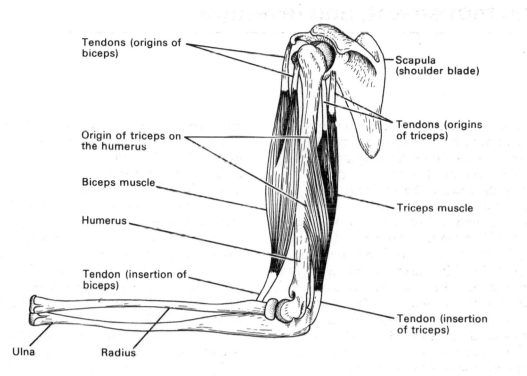

Fig. 2. Antagonistic muscles in the human arm. The biceps is the flexor muscle and the triceps is the extensor. Muscles are attached to bones by tendons. The tendon closest to the joint is called the insertion of the muscle and the tendon at the opposite end is the origin of the muscle. During muscular contraction the insertion moves while the origin remains fixed as the anchorage point of the muscle.

A Fulcrum in centre, effort and load at opposite ends

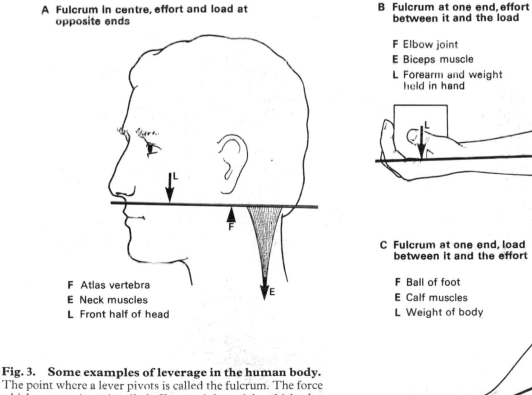

F Atlas vertebra
E Neck muscles
L Front half of head

B Fulcrum at one end, effort between it and the load

F Elbow joint
E Biceps muscle
L Forearm and weight held in hand

C Fulcrum at one end, load between it and the effort

F Ball of foot
E Calf muscles
L Weight of body

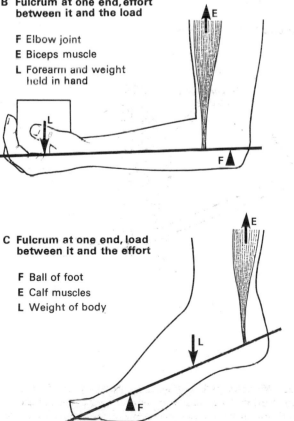

Fig. 3. Some examples of leverage in the human body. The point where a lever pivots is called the fulcrum. The force which moves a lever is called effort, and the weight which a lever moves is called the load. Name the fulcrum, effort, and load in each lever system illustrated above.

Sprains, dislocations, and fractures

Sprains

A sprain occurs when the ligaments of a joint are over-stretched and torn, causing swelling and pain. The commonest is an ankle sprain.

Treatment   Sprains will heal in about ten days if supported with pads and bandages, as illustrated in Figure 1.

Dislocations

A dislocation is the separation of bones at a joint. The commonest dislocations occur at the shoulder, elbow, thumbs, and fingers. The joint can no longer be moved, is very painful, and looks deformed (Fig. 3).

Treatment   Do not try to put the bones back in place. Support the limb in a comfortable position with a sling (Fig. 2), or a splint (Fig. 5). An arm can be bound to the body (Fig. 6), or the injured leg can be bound to the good leg in a lying position. Seek medical aid.

Fractures

A fracture is a broken bone. Figure 4 illustrates the commonest types of fracture. A fractured limb often has an odd shape and a painful swelling.

Treatment   Avoid moving the injured part because broken bones can severely damage muscles, blood vessels, and nerves. Cover any protruding bones with sterile dressing. Support the limb with a splint (Fig. 5), a sling (Fig. 2), or by binding it to the body (Fig. 6). The broken leg can be bound to the good leg. Seek medical aid.

A Surround the joint with a thick layer of cotton wool

B Bandage firmly as shown

C Apply a second layer of cotton wool

D Bandage again

Fig. 1.   Treatment of a sprained ankle. Sprains occur when ligaments are stretched and torn. They will heal in about ten days if supported with pads and bandages as shown, and rested.

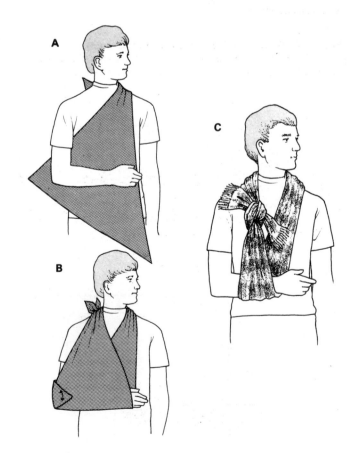

Fig. 2.   How to tie an arm sling. Fold a large piece of cloth into a triangle. **A** Put the cloth between the chest and forearm as shown. **B** Bring the bottom up in front of the arm and tie the points behind the neck. **C** Belts, scarves, neck ties, etc. can be used as a sling.

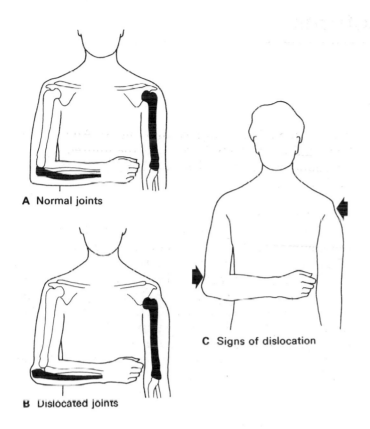

A Normal joints

B Dislocated joints

C Signs of dislocation

Fig. 3.   A dislocated joint cannot be moved, is very painful, and looks deformed. Do not try to put the bones back in place. Support the joint with a sling or splint.

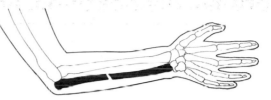

A Closed fracture The skin is not broken

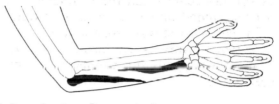

B Open fracture Bone protrudes or there is a deep wound over it

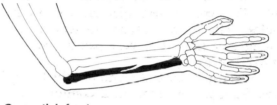

C Greenstick fracture The bone is bent or not completely broken

Fig. 4.   The commonest types of fracture. Do not move the injured part. Cover any exposed bone with sterile dressing. Support the limb with a sling or splint.

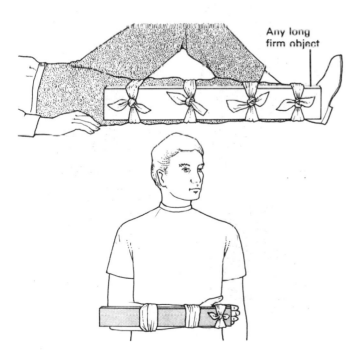

Any long firm object

Fig. 5.   The purpose of a splint is to immobilize a broken limb or dislocated joint until medical aid is available. Wood, metal, or even stiffly rolled paper can be used. Bind the splint to the limb at both sides of the break but not too tightly or blood circulation will be prevented.

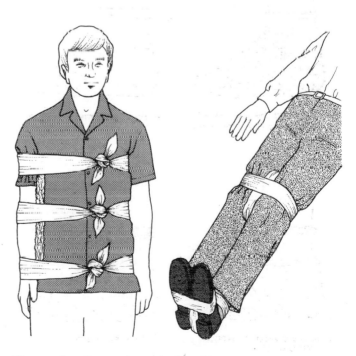

Fig. 6.   An alternative to splints. In 80 per cent of breaks, and in some dislocations, the body can be used as a splint. Bind the arm to the body using pads and bandages as shown. The broken leg can be bound to the good leg in the same way.

Muscles never completely relax. In every muscle there are always a few contracted fibres. These give rise to a slight muscular tension throughout the body which is called **muscle tone**.

Muscle tone has two functions. It keeps muscles ready for immediate contraction on demand, and it keeps the body in an upright position, or **posture**, without much conscious thought. If muscle tone were suddenly lost and all muscles relaxed completely, the body would collapse to the ground.

Upright posture is maintained by partial contraction of *both* the flexor and extensor muscles at certain joints. This holds the joints steady.

Good and bad posture

If good posture is achieved, the body is held upright with very little muscular effort. In a good standing posture, for example, the head is balanced on top of the backbone, the backbone is balanced on its point of attachment to the hip bone, the knee joints are straightened as far as their ligaments allow, and the feet are squarely on the ground (Fig. 1). In this position the muscles carry little weight; they simply keep the body balanced.

If the body is allowed to sag or stoop into a bad posture, the body is no longer balanced (Fig. 2). Muscles then carry far greater weight and quickly become tired and begin to ache. Organs such as the digestive system and main blood vessels are compressed and it is difficult to breathe properly. Prolonged bad posture can lead to permanent deformity, especially of the backbone.

Correcting bad posture

Habitual bad posture can be corrected by conscious effort and by regular exercise, since this improves muscle tone. Bad posture may also be caused by being overweight, and by poor working conditions such as dim lighting and low work surfaces. These should be avoided because they make it necessary to lean forwards for long periods. This puts the back and neck muscles under great strain, deforms the backbone, cramps the digestive system, and restricts breathing.

Poorly fitting clothing and shoes can also result in bad posture, and sometimes cause deformity of the bones (Fig. 3).

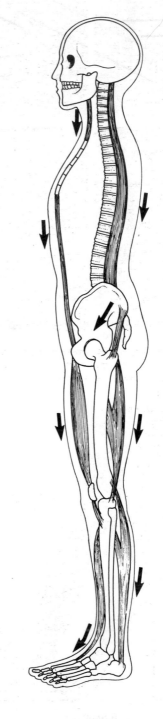

Fig. 1.   The muscles which keep the body upright. Each joint is held steady by contraction of the flexor and extensor muscles at the same time. In a good upright posture (shown above) the muscles carry very little weight. They simply keep the body balanced on its feet. The arrows indicate the direction of muscle pull.

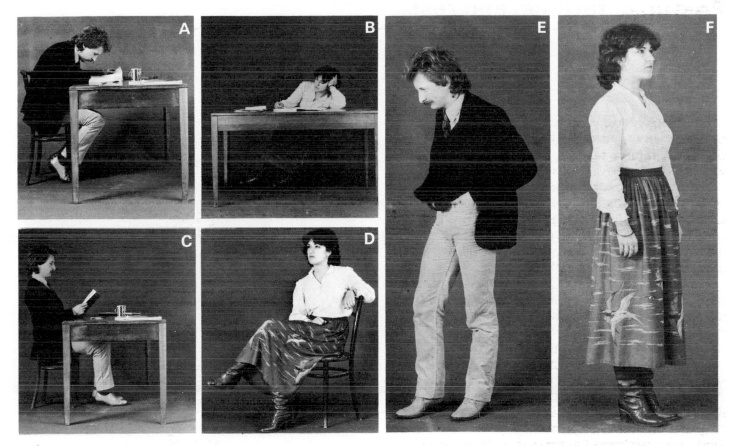

Fig. 2. Good and bad posture. Sort these photographs into examples of good and bad posture, then describe what is good and bad about each. If the body is allowed to take up a bad posture the backbone can bend out of shape, muscles can be strained and soon begin to ache, the digestive system can be compressed, and breathing movements can be restricted.

A Some effects of high heels

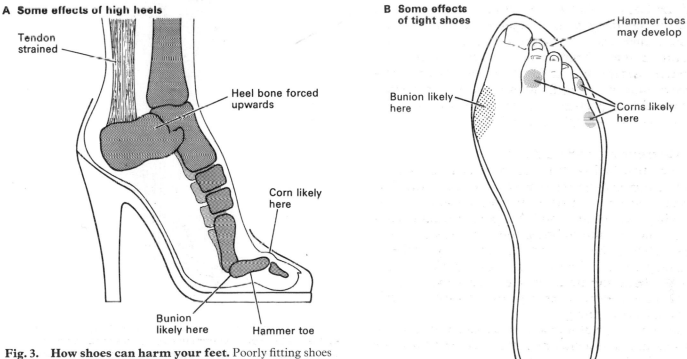

Tendon strained

Heel bone forced upwards

Corn likely here

Bunion likely here

Hammer toe

B Some effects of tight shoes

Hammer toes may develop

Bunion likely here

Corns likely here

Fig. 3. How shoes can harm your feet. Poorly fitting shoes can push bones out of their normal alignment; produce bunions, corns, and hammer toes; and force the body into a bad posture when standing.

18

Exercise, fatigue, and rest

Why exercise?

There are four main benefits to be gained from regular exercise.

1. *Suppleness* If joints are not regularly put through their full range of movements they tend to stiffen. Furthermore muscle tone, described in Unit 17, is reduced so that good posture is difficult to maintain. As a result, movement becomes restricted and even painful. In other words, suppleness has been reduced.

2. *Strength* Muscles waste away if not used regularly. On the other hand regular exercise maintains the efficiency, size, and strength of muscles.

3. *Stamina* If people who take little exercise are suddenly forced into some strenuous activity, they quickly become tired and out of breath and their pulse rises to an uncomfortably high level. Regular exercise increases the efficiency of heart muscles and of the muscles used for breathing, thereby increasing stamina. This means the heart and breathing mechanisms can respond quickly to sudden demands, and work hard for long periods.

4. *Health* In addition to increasing suppleness, strength, stamina, and muscle tone, regular exercise reduces the risk of heart disease, improves appetite, aids digestion, and leads to healthier hair and skin. Exercise is particularly important during the growth period because it influences the development of bones and muscular co-ordination.

Unfit people should start with a few repetitions of exercises illustrated in this Unit. People who get little exercise in their jobs should build up to five minutes of loosening-up daily (Fig. 1), then, in addition, a further twenty minutes three times a week on strength exercises (Fig. 2) and stamina exercises such as jogging, swimming, or cycling. A further two-hour walk at weekends would be very beneficial.

Fatigue, rest, and sleep

Muscular fatigue is relieved by relaxation and rest. Mental fatigue is often relieved by exercise, but rest is eventually necessary. Adequate sleep is essential for maintaining physical, and especially mental, health. There is no fixed rule as to the amount of sleep required for health, because sleep requirements vary enormously from person to person.

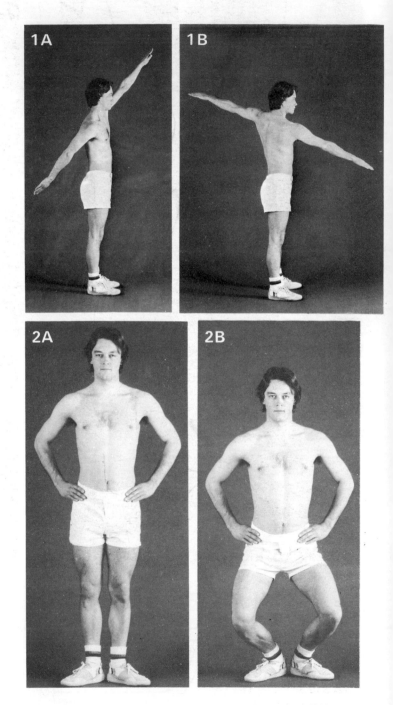

Fig. 1. **Exercises which develop suppleness** (also known as loosening-up exercises). These exercises put the main joints through their full range of movements, preventing them from becoming stiff through lack of use. They also help to maintain muscle tone. Repeat these exercises 10 times, progressing to 25.

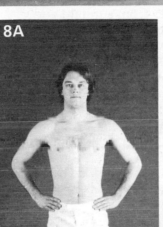

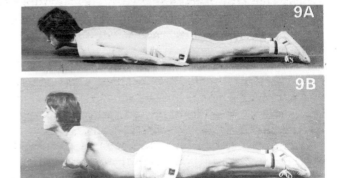

1 A and B Arm circling
2 A and B Half knee bending
3 A and B Trunk bending
4 A and B Trunk rotation
5 Knee raising This exercise should also be done in a standing position

6 A and B Press-ups
7 A and B Sit-ups
8 A and B Knee bends
9 A and B Back strengthener Raise head three times, then lift trunk up from the waist, with hands linked across chest, three times

Fig. 1 continued

Fig. 2. Exercises for strength. These exercises will maintain the size and strength of arm, leg, abdomen, and back muscles. Repeat these exercises 8 times, progressing to 12.

Photosynthesis and food chains

The food which humans and all other animals eat comes directly or indirectly from green plants. In order to understand why this is so it is necessary to study how plants obtain food. Plants, unlike animals, do not eat other living things. They make food by means of a process called photosynthesis.

Photosynthesis

This process occurs in leaves and other green parts of plants. It requires energy, carbon dioxide gas, and water. The energy comes from sunlight and is absorbed by a green substance called **chlorophyll**. Carbon dioxide comes from the air and enters plants through tiny pores called **stomata** which are mainly on leaves. Water is absorbed from the soil by the roots.

During photosynthesis sunlight energy is used to combine carbon dioxide with water. This produces glucose sugar, and oxygen gas is released into the air as a waste product. Photosynthesis can be summarized as follows:

$$6CO_2 + 6H_2O + Light \xrightarrow{\text{Chlorophyll}} C_6H_{12}O_6 + 6O_2$$

carbon dioxide   water   energy   glucose   oxygen

Plants use the glucose produced by photosynthesis to make all the carbohydrates, fats, and proteins which make up a plant body. These, in turn, make up the food of animals which eat plants. Animals which eat plants are called **herbivores** (e.g. grasshoppers and rabbits), and herbivores are eaten by **carnivores** (e.g. cats and foxes).

Plants use sunlight energy to make food. Some of this energy then passes to herbivores when they eat plants, and to carnivores when they eat herbivores (Fig. 1). This flow of energy from plants to herbivores and carnivores leads to the formation of food chains.

Food chains

All food chains begin with green plants. Plants are called the **producers** of the food chain because they make food by photosynthesis. All the other organisms in a food chain are called **consumers**. Herbivores are called **primary consumers** because they eat the producers (plants). Carnivores which eat herbivores are called **secondary consumers** and carnivores which eat secondary consumers are **tertiary consumers**.

Parasites (germs, tapeworms, etc.) feed on all members of a food chain. **Scavengers** (carrion crows, earthworms, etc.) and **decomposers** (certain fungi and bacteria) feed on the dead bodies of plants and animals which make up food chains. They break them down into substances like minerals and humus which fertilize the soil and are essential for healthy growth of plants and animals. Scavengers and decomposers are therefore responsible for passing these substances through food chains over and over again.

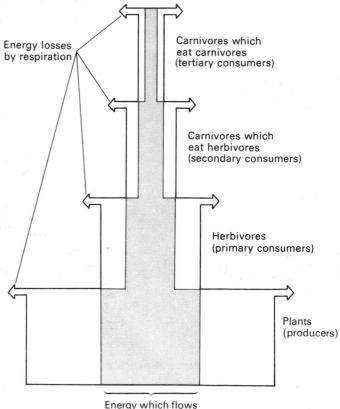

Energy losses by respiration

Carnivores which eat carnivores (tertiary consumers)

Carnivores which eat herbivores (secondary consumers)

Herbivores (primary consumers)

Plants (producers)

Energy which flows along food chain

Fig. 1. Flow of energy through a food chain. Plants use sunlight energy to make food by photosynthesis. Some of this energy passes to herbivores when they eat plants and then to carnivores when they eat herbivores. This flow of energy from plants to herbivores and carnivores is called a food chain.

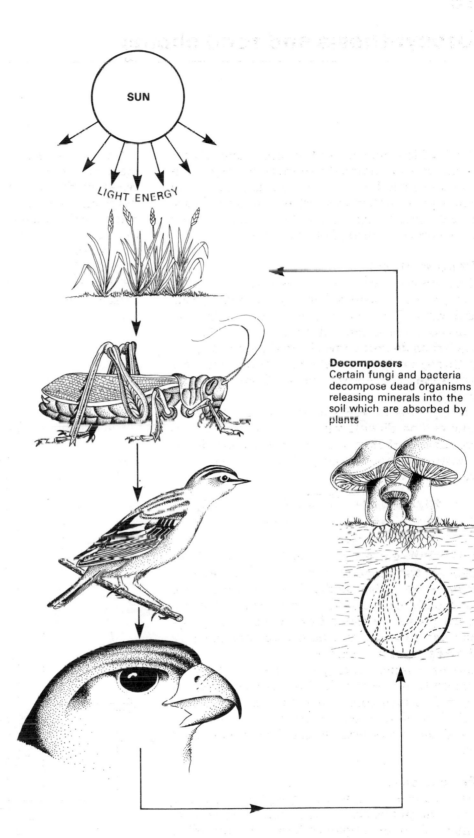

Sunlight
The source of energy for photosynthesis

SUN

LIGHT ENERGY

Producers
Green plants (e.g. grasses) make food by photosynthesis

Primary consumers
Herbivores (e.g. grasshoppers) eat plants

Secondary consumers
Carnivores (e.g. warbler) eat herbivores

Tertiary consumers
Carnivores (e.g. hawks and kestrels) eat secondary consumers

Decomposers
Certain fungi and bacteria decompose dead organisms releasing minerals into the soil which are absorbed by plants

Fig. 2. A simple food chain. All food chains begin with green plants, which are called producers because they make food. All other organisms in a food chain are consumers. Decomposers break down bodies of dead plants and animals into substances which are absorbed by plants and pass back into the food chain.

39

Types of food and their energy value

Food provides the body with energy, and with the raw materials for growth and repair of its tissue. Energy is released from food during the chemical reactions of respiration, described in Unit 5. The amount of energy which the body can obtain from a food (the food's energy value) is now measured in units called kilojoules (abbreviated to kJ). Calories are no longer used in scientific work (1000 calories (1 kcal) = 1 Calorie = 4·2 kJ). The energy values of some foods are given in the table opposite.

Carbohydrates

Carbohydrates are sugary and starchy foods. Carbohydrates are the body's main source of energy (1 gram of carbohydrate provides 17 kJ of energy). If too much carbohydrate food is eaten the body converts it into fat and stores it under the skin and around various organs.

Fats and oils

Fats and oils are the body's main stored foods. When they are eventually respired 1 gram of fat provides 39 kJ of energy. Animal fats are more easily digested than vegetable fats and contain more vitamin A and D. A layer of fat under the skin helps reduce the amount of heat lost from the body in cold weather. The fat surrounding body organs helps keep them in their correct position.

Proteins

Proteins are used for growth, and for replacing and repairing worn out and damaged tissues. Protein foods are needed particularly by growing children, pregnant women, and people recovering from injuries or sickness. Proteins can provide 17 kJ of energy per gram.

During digestion proteins are broken down into chemicals called **amino acids**. There are about twenty-six amino acids but only ten of them are essential to human health. Every type of animal protein contains all the essential amino acids, and so they are called **first-class proteins**. No one plant protein contains all the essential amino acids but all of them can be obtained by eating a wide variety of plant foods. Therefore plants are said to contain **second-class proteins**.

Fig. 1. Types of food. Use the table opposite to sort these foodstuffs into four groups: rich in protein, rich in fat, rich in carbohydrate, and mixed foods (that is, containing large amounts of two or more types of food).

Table 1

Composition of food per 100 g (percentage not accounted for is inedible waste, e.g. shell, bone, skin, water, etc.). Figures are taken from the *Manual of Nutrition*, Ministry of Agriculture, Fisheries and Food, H.M. Stationery Office, 1976. This chart provides the information required to answer the comprehension questions on page 187.

Type of food	Kilojoules	Proteins g	Fats g	Carbo-hydrates g
Meat				
Bacon, grilled	1852	24·5	38·8	0
Beef, roast	932	30·9	11·0	0
Chicken, roast	621	24·8	5·4	0
Liver, fried	1020	24·9	13·7	5·6
Luncheon meat	1298	12·6	26·9	5·5
Pork chop, grilled	1380	28·5	24·2	0
Sausage, beef	1242	9·6	24·1	11·7
Dairy produce				
Butter	3006	0·5	81·0	0
Cheese	1708	25·4	34·5	0
Eggs, raw, one	612	12·3	10·9	0
Ice-cream	805	4·1	11·3	19·8
Milk, liquid whole	274	3·3	3·8	4·8
Yoghurt, fruit	410	4·8	1·0	18·2
Fish				
Cod, fried in batter	834	19·6	10·3	7·5
Kipper	770	19·8	11·7	0
Sardines, canned	906	23·7	13·6	0
Cereals				
Bread, white	1068	8·0	1·7	54·3
Bread, wholemeal	1025	9·6	3·1	46·7
Rice, steamed	1531	6·2	1·0	86·8
Vegetables				
Beans, canned	266	5·1	0·4	10·3
Brussel sprouts, cooked	75	2·8	0	1·7
Cabbage, cooked	66	1·7	0	2·3
Carrots, cooked	98	0·7	0	5·4
Lettuce	36	1·0	0	1·2
Peas, cooked	208	5·0	0	7·7
Potatoes, boiled	339	1·4	0	19·7
Potato chips, fried	1028	3·8	9·0	37·3
Tomatoes	52	0·8	0	2·4
Fruit				
Apples	197	0·3	0	12·0
Bananas	326	1·1	0	19·2
Oranges, peeled	150	0·8	0	8·5
Plums	137	0·6	0	7·9
Strawberries	109	0·6	0	6·2
Miscellaneous				
Apple pie	1179	3·2	14·4	40·4
Buns, currant	1385	7·8	8·5	58·6
Coffee, white, 1 cup	84	0	0·5	0·5
Fruit cake, rich	1546	4·6	15·9	55·0
Rice pudding	594	3·6	7·6	15·7
Sugar, white	1654	0	0	100·0
Tea with milk, 1 cup	84	0	0·5	0·5

21

Vitamins and minerals

Vitamins are chemicals which are essential for growth and general health. The body needs only very small amounts of vitamins, but if they are missing from the diet illnesses called **vitamin deficiency diseases** develop.

Vitamin A keeps the skin and bones healthy, helps prevent infection of the nose and throat, and is necessary for vision in dim light. Lack of vitamin A causes poor night vision, and increases the chances of infection of the nose and throat. Vitamin A is found in carrots, milk, fish-liver oils, and green vegetables.

Vitamin B$_1$ helps the body obtain energy from food. Lack of it reduces growth, and causes **beri-beri**, a disease in which the limbs are paralysed. Vitamin B$_1$ is found in yeast, wholemeal bread, nuts, peas, and beans.

Vitamin B$_2$ enables the body to obtain energy from food. Lack of it causes stunted growth, cracks in the skin around the mouth, an inflamed tongue, and damage to the cornea of the eye. Vitamin B$_2$ is found in liver, milk, eggs, yeast, cheese, and green vegetables.

Vitamin B$_{12}$ enables the body to form protein and fat, and to store carbohydrate. Lack of it causes **pernicious anaemia** (failure to produce haemoglobin for red blood cells). Vitamin B$_{12}$ is found in liver, meat, eggs, milk, and fish.

Vitamin C is destroyed by cooking, grating, or mincing food. It disappears from food if it is stored for long periods. Vitamin C helps wounds to heal, and is needed for healthy gums and teeth. Lack of it causes **scurvy**, a disease in which the gums become soft, the teeth grow loose, and wounds fail to heal properly. Vitamin C is found in oranges, lemons, black currants, green vegetables, tomatoes, and potatoes.

Vitamin D enables the body to absorb calcium and phosphorus from food. These chemicals are needed to make bones and teeth. Lack of vitamin D causes **rickets** (soft weak bones, which bend under pressure). Vitamin D is found in liver, butter, cheese, eggs, and fish.

Vitamin K is needed to make blood clot in wounds. Lack of it can cause **haemorrhage** (excessive bleeding) whenever the skin is broken. Vitamin K is obtained from cabbage and cereals, and is made by bacteria which live inside the digestive system.

Minerals   Humans require about fifteen different minerals in their diet. Most of these are supplied by meat, eggs, milk, green vegetables, and fruit.

Mineral	Daily requirement in mg	Functions in the body
Sodium chloride (common salt)	5–10	Blood plasma is almost 1% salt. Salt is also needed for digestion, and to enable impulses to pass along nerve fibres.
Potassium	2	Needed for muscular contraction.
Magnesium	0·3	
Phosphorus	1·5	Forms a large part of bones and teeth. Also needed for chemical reactions of respiration.
Calcium	0·8	Calcium salts form an important part of bones and teeth.
Iron	0·01	A large part of haemoglobin is made up of iron. Haemoglobin is the substance which gives blood its red colour. It transports oxygen to the tissues.
Iodine	0·00003	Needed by the thyroid gland, which produces hormones. A lack of iodine causes the disease goitre, in which the thyroid gland becomes enlarged.

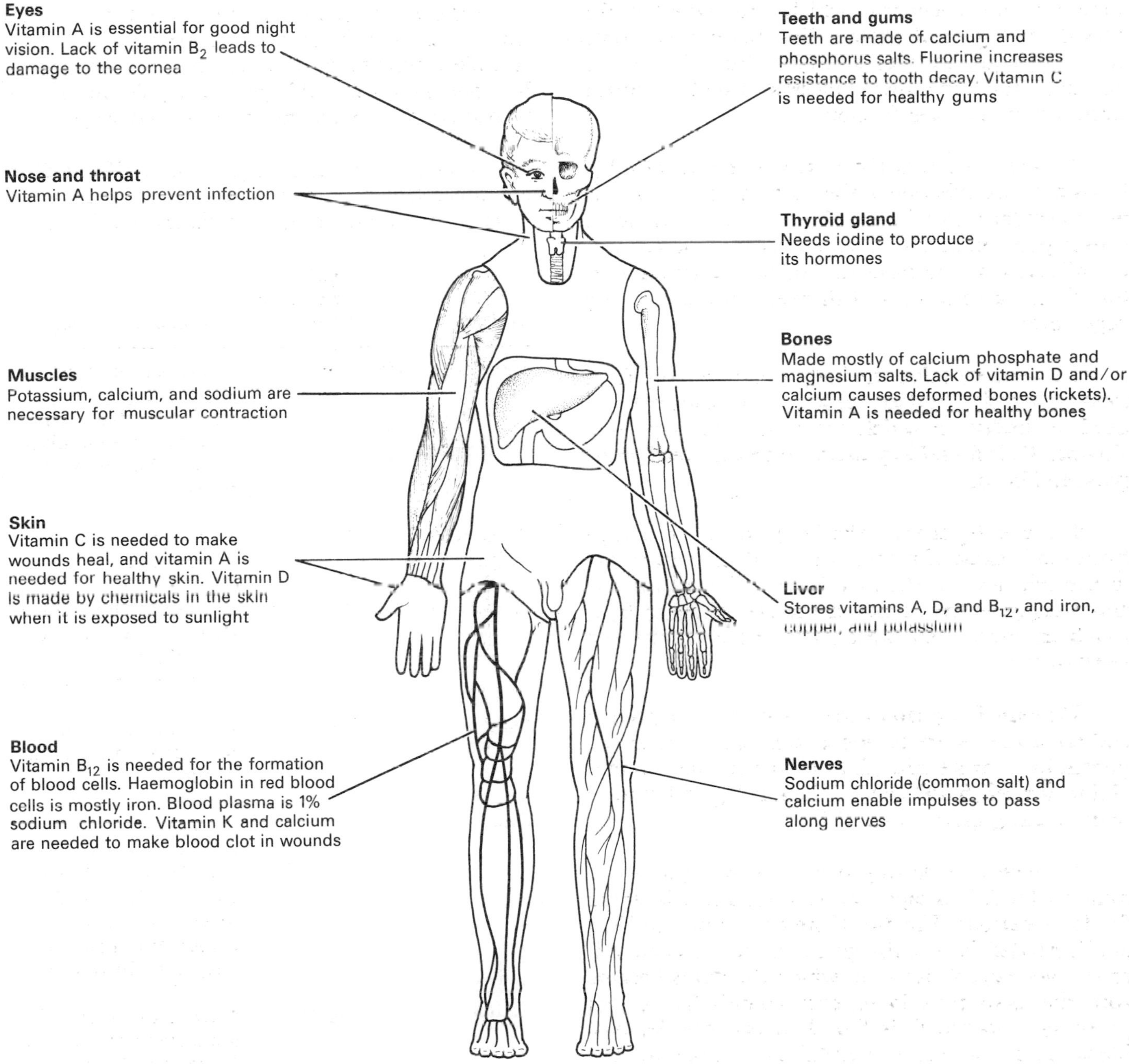

Eyes
Vitamin A is essential for good night vision. Lack of vitamin B_2 leads to damage to the cornea

Nose and throat
Vitamin A helps prevent infection

Muscles
Potassium, calcium, and sodium are necessary for muscular contraction

Skin
Vitamin C is needed to make wounds heal, and vitamin A is needed for healthy skin. Vitamin D is made by chemicals in the skin when it is exposed to sunlight

Blood
Vitamin B_{12} is needed for the formation of blood cells. Haemoglobin in red blood cells is mostly iron. Blood plasma is 1% sodium chloride. Vitamin K and calcium are needed to make blood clot in wounds

Teeth and gums
Teeth are made of calcium and phosphorus salts. Fluorine increases resistance to tooth decay. Vitamin C is needed for healthy gums

Thyroid gland
Needs iodine to produce its hormones

Bones
Made mostly of calcium phosphate and magnesium salts. Lack of vitamin D and/or calcium causes deformed bones (rickets). Vitamin A is needed for healthy bones

Liver
Stores vitamins A, D, and B_{12}, and iron, copper, and potassium

Nerves
Sodium chloride (common salt) and calcium enable impulses to pass along nerves

Fig. 1.   Summary of the functions of the main vitamins and minerals required by humans.

43

To a large extent good health depends upon eating the correct *amounts* of food and the correct *types* of food. The amount of food which a person requires depends upon the energy used each day. Types of food should be chosen which make up a balanced diet, as explained below and in Figure 1.

Food, and daily energy requirements

The body requires a certain amount of energy each day to function properly and this energy comes from food. The total amount of energy used each day varies according to age, sex, body size, occupation, and special conditions such as pregnancy (see table opposite). Ideally the food eaten each day should supply no more and no less than the amount of energy used in that day. If, over a long period, insufficient food is eaten to supply this energy the body uses its stored fat and then protein and soon becomes thin and unhealthy. If excess food is eaten it is converted into fat and the body becomes overweight.

A balanced diet

It is unhealthy to supply the body's energy needs by eating only one type of food, such as carbohydrate, and little else. A balanced diet is more healthy: the energy-giving foods (fats and carbohydrates) in each meal are 'balanced' by amounts of body-building foods (proteins) and protective foods (vitamins and minerals) (Fig. 1). A balanced meal consists of about one part protein, one part fat, and four or five parts carbohydrate, and it includes vitamins and mineral-rich foods.

Proteins 50 to 60 per cent of the protein eaten daily should consist of first-class proteins. If these are not available a wide variety of second-class proteins should be eaten, as explained in Unit 20.

Fats and carbohydrates A small amount of fat is required to supply the fat-soluble vitamins A, D, and E, and essential fatty acids. Beyond these needs fat is not essential but, when used for frying and baking, it makes food more tasty. Fat also reduces the bulk of food needed to supply energy requirements since 1 gram of fat contains double the energy of 1 gram of carbohydrate. Eating large amounts of fat, however, is dangerous to the health, as explained in Unit 73. When choosing carbohydrate foods it is important to avoid large amounts of sugar as this increases tooth decay.

In a balanced diet the daily intake of:

Energy-rich foods
(fats, oils, and carbohydrates)

Body-building foods
(proteins)

Protective foods
(vitamins and minerals)

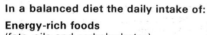
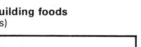

total daily energy
requirements of
the body

Fig. 1. **A balanced diet** is one in which carbohydrate foods are 'balanced' by proteins, vitamins, and minerals. In addition, as illustrated above, the amount of food eaten each day should supply no more and no less than the amount of energy used that day.

Table 2. Variation in the amount of energy which the body uses each day, according to age, sex, occupation, and special conditions such as pregnancy, together with notes on diet.

Figures are from a Department of Health and Social Security pamphlet (1969)

Age, sex, occupation, etc.	Energy used in 1 day		Diet
Birth to 1 year	0–3 months 2300 kJ 3–6 months 3200 kJ 6–9 months 3800 kJ 9–12 months 4200 kJ		At least 2 g of protein are needed per kg of body weight from 0–6 months and 1·6 g per kg from 6–9 months, gradually reducing to 1 g per kg. Weaning usually begins at about 4 months
8 years	**Active children** 8800 kJ (both sexes)		At least 30 g of protein a day but 53 g are recommended
15 years	**Males** 12 600 kJ	**Females** 9600 kJ	
Adult (light work)	11 550 kJ	9450 kJ	80–100 g of protein per day and 60% of this should be first-class protein
Adult (moderate work)	12 100 kJ	10 500 kJ	About 300 g of carbohydrate per day except for those doing heavy work who should eat far more Sugar should not be eaten in large quantities as it increases tooth decay About 100 g of fat should be eaten per day
Adult (heavy work)	15 000 kJ to 20 000 kJ	12 600 kJ	
Pregnant and nursing mothers	**Pregnant** 10 000 kJ	**Nursing** 11 300 kJ	Pregnant women should eat about 85 g of protein per day increasing to 100 g during breast feeding, together with increased amounts of food containing calcium, iron, and all vitamins

Food preservation

If food is not quickly preserved in some way after it is harvested it soon 'goes bad' or decomposes, owing to the action of microbes such as bacteria and fungi which begin to grow and multiply in it. Apart from giving food an unpleasant taste, these microbes can be a danger to health. Bacteria of the genera *Clostridium* and *Salmonella* release poisons (toxins) into the food in which they live. These poisons can cause serious illness or death, and can reach dangerous levels even before the food develops an unpleasant taste or smell.

Preserved food can be transported long distances from where it is produced, and stored for long periods in warehouses or homes without risk of its decomposing or endangering health.

To live in food microbes require moisture, warmth, and usually oxygen. Food can be preserved by storing it in conditions where these requirements are absent.

Drying or dehydration
This is one of the oldest methods of food preservation. Hunters of long ago used to dry strips of meat or fish in the sun, and in parts of Africa this method of preservation is still used. Grapes can be dried to form raisins, sultanas, and currants, and eggs and milk can be dried to a powder and stored indefinitely. Food can now be dried without heat by placing it in a vacuum for a little while. The advantage of vacuum drying is that, unlike drying in heat, it does not alter the flavour of food or destroy any of its vitamins.

Freezing
Domestic refrigerators keep food at about 4°C. At this temperature microbes reproduce very slowly and so food remains fresh for a few days. At temperatures below freezing point bacteria and fungi are unable to decompose food or multiply. A domestic deep-freeze keeps food at −20°C or below. At this temperature food remains fresh for several months. It is essential to cook food soon after it has thawed out as any microbes in it will quickly resume multiplying.

Canning and bottling
Canning is one of the safest methods of preserving food for long periods. The food is first heated to a temperature which kills microbes. In other words, it is **sterilized**. Then, while still hot, the food is put into sterilized cans. Further heating drives all the air out of the can before it is sealed. Domestic bottling of food is done in a similar way, but it is more difficult to seal a bottle so that it remains airtight for long periods.

Chemical preservatives
Salting This is a very old method of preserving certain types of fish and meat. The food is placed in common salt (sodium chloride). The salt draws most of the water from the food, and it kills any microbes present by drawing water out of them also. Salted butter will keep fresh longer than unsalted.

Sugar Like salt, strong sugar solution kills microbes by drawing water from them. It is sugar which keeps jam fresh for long periods.

Smoking Certain types of wood smoke contain chemicals called **phenols**. Phenols poison microbes, and so food can be preserved by hanging it in a smoky atmosphere until sufficient phenols have been absorbed. Smoked (kippered) herring, smoked haddock, and smoked bacon are prepared in this way.

Other chemical preservatives Onions, cabbage, cauliflower, and gherkins can be preserved (pickled) in vinegar. Wine is simply fruit juice preserved in ethyl alcohol. Light wines and beer contain insufficient alcohol to preserve them and so sulphurous acid is added to prevent decomposition. Benzoic acid, boric acid, salicylic acid, and sulphur dioxide are often used, in very small quantities, to preserve meat products such as sausage.

Pasteurization of milk
The French biologist Louis Pasteur discovered that bacteria can be killed by strong heat. But, unfortunately, the high temperature needed to kill all bacteria alters the taste of food such as milk and also destroys certain vitamins. If, however, milk is heated to only 62°C for thirty minutes and then quickly cooled most of the microbes present are killed including *Mycobacterium tuberculosis*, the cause of T.B. This process is called **Pasteurization**. Pasteurized milk will keep fresh for a day or two in cool conditions provided it is kept in sealed, sterilized bottles.

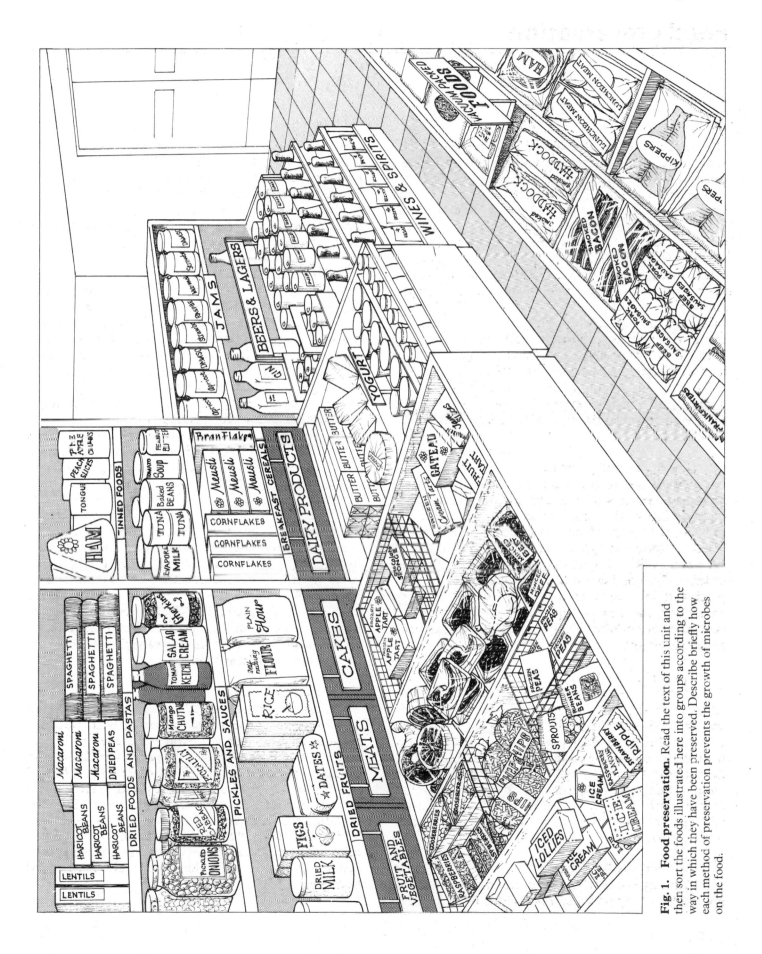

Fig. 1. Food preservation. Read the text of this unit and then sort the foods illustrated here into groups according to the way in which they have been preserved. Describe briefly how each method of preservation prevents the growth of microbes on the food.

Digestion, absorption, and assimilation

Digestion

Digestion is the process which makes food soluble. All foods must be made soluble before they can pass into the bloodstream and be transported around the body to all the cells.

Food is digested inside a tube called the **alimentary canal**. This tube starts at the mouth and ends at the anus. First the food is broken down into small pieces by the teeth. This is called mechanical breakdown of food. Next, chemical breakdown of food occurs as follows. In certain places the walls of the alimentary canal produce digestive juices containing chemicals called **digestive enzymes**. These enzymes break food down into chemically simpler substances which are soluble. One theory of how this happens is explained in Unit 4.

Digestive enzymes can be divided into groups according to the type of food they digest. **Amylases** are a group of enzymes which digest starchy foods into sugars such as glucose. **Lipases** are enzymes which digest fats and oils into simpler substances called **fatty acids** and **glycerol**. **Proteases** digest proteins into simpler substances called **amino acids**

(Fig. 1). When digestion is complete absorption takes place.

Absorption

Absorption is the movement of digested (soluble) foods through the walls of the alimentary canal into the bloodstream. The food is transported first to the liver and then to all the other cells of the body, where it is assimilated.

Assimilation

Assimilation is the name for the processes by which cells take in and make use of digested food to produce energy, and as a raw material for growth and repair of tissues—in other words, for metabolism.

Most foods contain substances which cannot be digested. Humans, for instance, have no enzymes which can digest the cellulose cell walls in foods taken from plants. Cellulose and other indigestible materials are called **faecal matter**, or **faeces**. Faeces pass out of the body through the anus. This process is called **defecation** (Fig. 2).

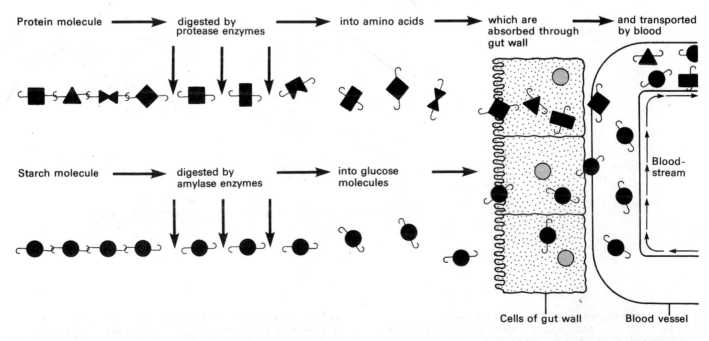

Fig. 1. Digestive enzymes digest food by breaking it down into chemicals which are soluble. Food then passes through the gut wall into the bloodstream—a process called absorption.

Blood transports food to all parts of the body where it is used (assimilated) by cells.

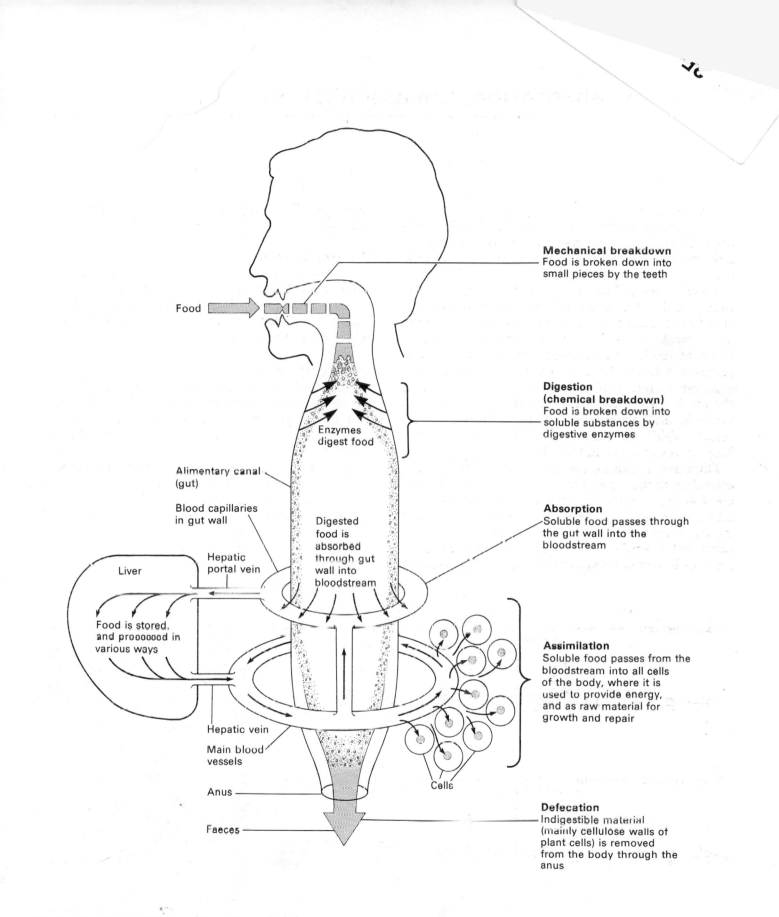

Mechanical breakdown
Food is broken down into small pieces by the teeth

**Digestion
(chemical breakdown)**
Food is broken down into soluble substances by digestive enzymes

Absorption
Soluble food passes through the gut wall into the bloodstream

Assimilation
Soluble food passes from the bloodstream into all cells of the body, where it is used to provide energy, and as raw material for growth and repair

Defecation
Indigestible material (mainly cellulose walls of plant cells) is removed from the body through the anus

Food

Enzymes digest food

Alimentary canal (gut)

Blood capillaries in gut wall

Hepatic portal vein

Liver

Food is stored and proooooed in various ways

Digested food is absorbed through gut wall into bloodstream

Hepatic vein

Main blood vessels

Anus

Faeces

Cells

Fig. 2. Food is digested inside a tube called the alimentary canal, or gut—shown here as a straight tube from the mouth to the anus. First, food is broken into small pieces by the teeth. Second, enzymes break down food into soluble chemicals. Third, the soluble food is absorbed through the gut wall into the bloodstream. Fourth, food is transported to the liver and then to all parts of the body in the blood. Fifth, the food is assimilated—it is used in the cells to provide energy, and as material for growth and repair.

Teeth

Structure of a tooth

Teeth grow in sockets in the jaw bones. Each tooth consists of a **root**, which is attached to the jaw bone by **cement** and tough fibres; a **neck**, which is the part surrounded by gum; and a **crown**, the part above gum level (Fig. 1). The crown is the biting surface. It has an outer layer of **enamel** which is the hardest substance in the body. The remainder of a tooth consists of a bone-like material called **dentine**, except for the central **pulp cavity** which contains nerves and blood vessels. Figure 2 illustrates the arrangement and types of teeth.

Milk and permanent teeth

Teeth develop inside the jaws before birth and begin to appear, or erupt, at about five months of age. By about age six years most children have twenty-four teeth. Twenty of these are **milk teeth**. These are shed between the ages of seven and eleven years and are replaced by larger **permanent teeth** (Fig. 3). By eighteen years of age most people have a total of thirty-two teeth. The last to appear are the rear molars, or wisdom teeth.

Care of teeth

Decay occurs because bacteria in the mouth change carbohydrate food stuck to teeth into acids which dissolve away first the enamel and then the dentine (Fig. 4A). Tooth decay can be avoided by brushing at the very least once a day in a way which removes all the food, especially that between the teeth. A side-to-side and up-and-down scrubbing action with the brush is *not* recommended because it leaves food between the teeth and can damage gums. Figure 4B illustrates one of the commonly recommended methods for thoroughly cleaning teeth.

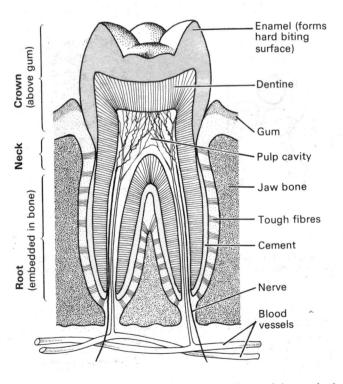

Fig. 1. Structure of a tooth. The root of a tooth is attached to the jaw bone by cement and tough fibres. The crown is the biting surface of a tooth, and is made of enamel. The inner part of a tooth is made of bone-like material called dentine, and contains a pulp cavity filled with nerves and blood vessels.

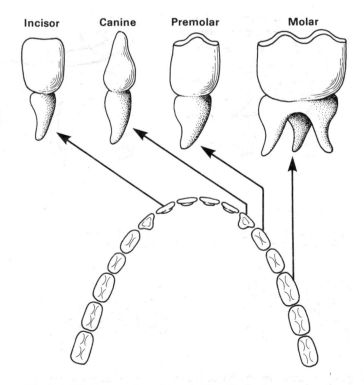

Fig. 2. There are four types of tooth. Incisors and canines are used for biting. Premolars and molars are used for chewing, which crushes and pulverizes food making it ready for digestion.

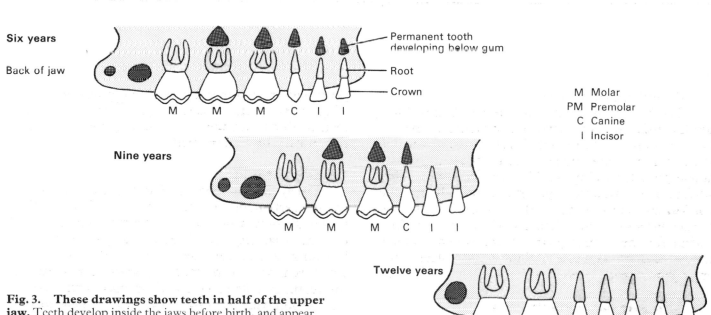

Six years

Back of jaw

Permanent tooth developing below gum

Root

Crown

M M M C I I

M Molar
PM Premolar
C Canine
I Incisor

Nine years

M M M C I I

Twelve years

M M PM PM C I I

Fig. 3.   These drawings show teeth in half of the upper jaw. Teeth develop inside the jaws before birth, and appear about five months after birth. Milk teeth are later replaced by permanent teeth. Find milk teeth in the drawings by looking for those which have a permanent tooth developing above them.

A Decay of a tooth

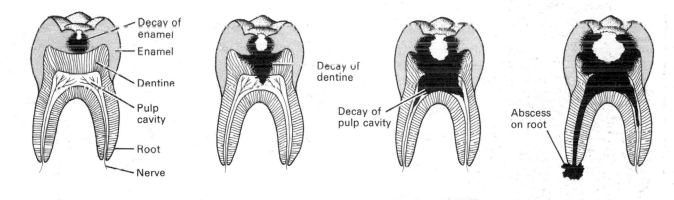

Decay of enamel

Enamel

Dentine

Pulp cavity

Root

Nerve

Decay of dentine

Decay of pulp cavity

Abscess on root

B Care of teeth

Downward strokes clean the upper teeth

Upward strokes clean the lower teeth

A back-and-forth action cleans the molars

An up-and-down action cleans behind the teeth

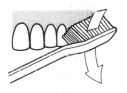
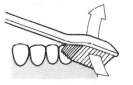
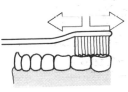
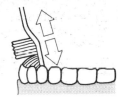

Fig. 4.   Care of teeth. Bacteria in the mouth digest food stuck to the teeth, changing it to acid which dissolves enamel and dentine. Brush the teeth as shown above at least once a day to remove all the food and bacteria on and inbetween the teeth.

Digestion

Digestion in the mouth

Chewing mixes food with **saliva**, produced by the salivary glands. Saliva moistens food so that it can be swallowed easily and it contains an enzyme, **salivary amylase**, which breaks down cooked starch into a sugar called maltose. During swallowing food is formed into a ball, or **bolus**, by the tongue and pushed to the back of the mouth. The soft palate is pushed upwards preventing food from entering the nasal cavity and the entrance to the wind-pipe is also closed (Fig. 1). The bolus then passes down the throat into the **oesophagus**, or gullet. Muscular contractions called **peristalsis**, described in Fig. 2, force food down into the stomach, and throughout the remainder of the digestive system.

Digestion in the stomach

The stomach has rings of muscle at both ends called **sphincters** (Fig. 4). After a meal both sphincters contract and keep food in the stomach for about an hour. Muscles in the stomach wall contract rhythmically, mixing food with a liquid called **gastric juice.** This contains **pepsin**, an enzyme which digests protein, and a weak solution of hydrochloric acid, which kills germs and helps the enzyme to work. When the lower sphincter opens, partly digested food moves into the **duodenum**, which is the first part of the **small intestine**.

Digestion in the small intestine

In the duodenum food is mixed with **bile**. Bile is made in the liver, which is described in Unit 28. Bile helps digestion by changing fats and oils into an **emulsion**, which means they are broken down into minute droplets. Bile also neutralizes stomach acid. This enables enzymes in the small intestine to work. Food is mixed with enzymes from the pancreas, and later with more enzymes from the ileum wall (Fig. 4). These enzymes digest food completely, which is then absorbed into the bloodstream.

Functions of the large intestine

The substances in food which cannot be digested pass into the large intestine. Here, water and salt are removed. The remaining faecal matter passes out of the body through the anus.

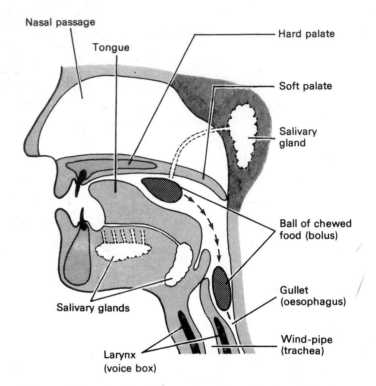

Fig. 1. Digestion in the mouth and swallowing. Chewing mixes food with saliva from the salivary glands. Saliva moistens food and contains an enzyme which breaks down cooked starch into maltose sugar. During swallowing the tongue pushes food to the back of the mouth from where it passes into the gullet. The soft palate prevents food from entering the nasal cavity and the entrance to the wind-pipe is also closed.

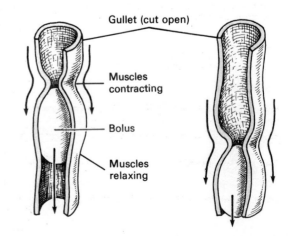

Fig. 2. Waves of muscular contraction called peristalsis move food along the gut. Circular muscles in the gut wall contract behind the food and relax in front of it, pushing it forwards at about 20 cm a second.

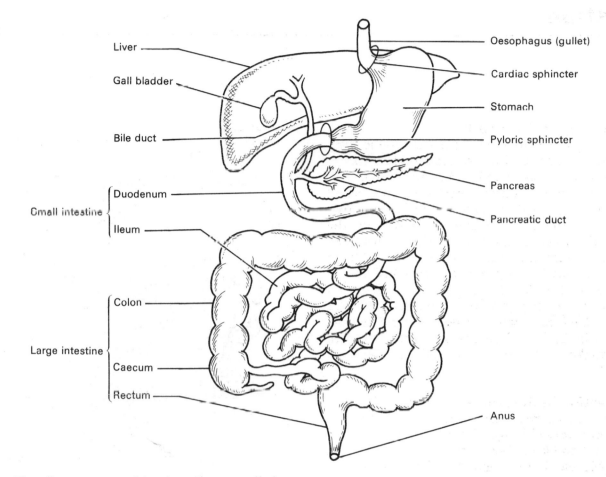

Fig. 3. The alimentary canal (gut) partly unravelled to show the pancreas.

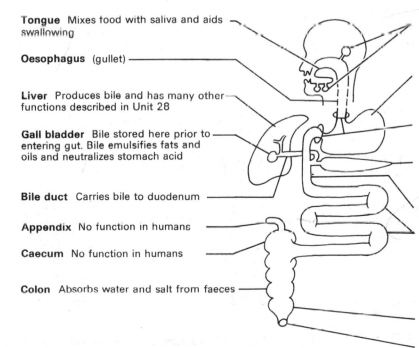

Tongue Mixes food with saliva and aids swallowing

Oesophagus (gullet)

Liver Produces bile and has many other functions described in Unit 28

Gall bladder Bile stored here prior to entering gut. Bile emulsifies fats and oils and neutralizes stomach acid

Bile duct Carries bile to duodenum

Appendix No function in humans

Caecum No function in humans

Colon Absorbs water and salt from faeces

Salivary glands Produce saliva, which moistens food and contains an enzyme, salivary amylase, which digests starch

Stomach A bag made of muscle which holds food for about an hour. Produces gastric juice containing hydrochloric acid, and pepsin which digests protein

Sphincters Rings of muscle which close off the stomach after a meal

Pancreas A gland which produces enzymes: trypsin, which digests proteins; amylase, which continues starch digestion; and lipase, which digests fats and oils

Duodenum Food mixed with bile

Ileum The ileum wall produces erepsin, which continues protein digestion; lipase, which continues fat and oil digestion; maltase, which digests maltose; sucrase, which digests sucrose; and lactase, which digests lactose

Rectum Holds indigestible matter prior to defecation

Anus Indigestible matter removed from the body (defecation)

Fig. 4. Diagram of the gut and functions of its parts.

Absorption

Digestive enzymes break down food into a liquid called **chyle**. This liquid is a mixture of many substances. It contains glucose and other sugars produced by digestion of carbohydrates; amino acids produced by protein digestion; fatty acids and glycerol from fat and oil digestion; and some undigested but emulsified fats. Chyle also contains vitamins and minerals. Nearly all absorption of this digested food takes place in the ileum of the small intestine.

The whole of the ileum's internal surface is covered by finger-like projections called **villi** (*singular*: villus) which are about 1 mm long (Figs. 1 and 2). Each square millimetre of ileum wall has up to forty villi and there are about five million in the ileum as a whole. The presence of villi gives the ileum a far greater internal surface area available for absorption than if it had a smooth lining. Originally it was thought that this area was about ten square metres in humans, but under very high magnification (× 40 000) the surface of each individual cell lining the ileum is seen to be folded into **microvilli** (Fig. 2B), giving the ileum an estimated total internal surface area of thirty square metres.

Inside each villus there is a dense network of blood capillaries, and a single **lacteal**, or lymph vessel, which is closed at its upper end (Figs. 1 and 2B). After digested food has been absorbed into the villus, it passes into these two kinds of vessels.

Amino acids, sugars, vitamins, minerals, and small amounts of fatty acids and glycerol pass into the blood capillaries of the villi. These capillaries join together to form a large blood vessel called the hepatic portal vein which carries the food to the liver. What happens to this food next is described in Unit 28. The bulk of the fatty acids and glycerol pass through the villi walls into the lacteals. The remaining emulsion of oil droplets is also absorbed into the lacteals of the villi. This happens in the following manner. The oil droplets pass between the microvilli of the cells covering each villus and are absorbed whole into these cells. The droplets then pass through the cytoplasm of the cells and out of the other side into the lacteals. These oil droplets, together with more droplets built up from absorbed fatty acids and glycerol, pass into the main lymphatic system and are eventually discharged into the bloodstream, as described in Unit 34.

Movement of absorbed food out of the villi into the bloodstream and lymphatic system is assisted by periodic contractions of villi, during which they suddenly become shorter and fatter and then relax slowly. This happens about six times a minute in each villus.

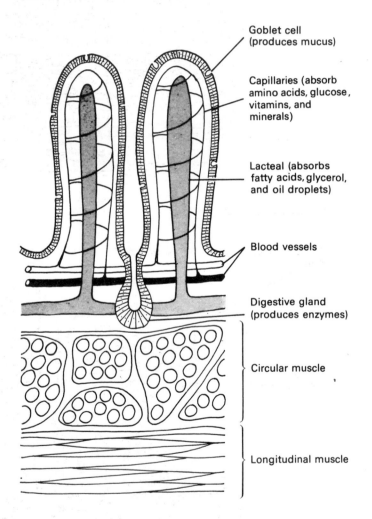

Goblet cell (produces mucus)

Capillaries (absorb amino acids, glucose, vitamins, and minerals)

Lacteal (absorbs fatty acids, glycerol, and oil droplets)

Blood vessels

Digestive gland (produces enzymes)

Circular muscle

Longitudinal muscle

Fig. 1.   Diagram of the ileum wall. Soluble (digested) food is absorbed in the ileum, through the surface of finger-like projections called villi. The presence of villi gives the ileum a far greater internal surface area for absorption than if it had a smooth lining. Amino acids, sugars, vitamins, and minerals are absorbed into blood capillaries inside each villus. The blood carries these foods to the liver and then distributes them around the body. Fatty acids, glycerol, and tiny oil droplets are absorbed into a lacteal at the centre of each villus. Then they are transported through the lymphatic system and into the bloodstream.

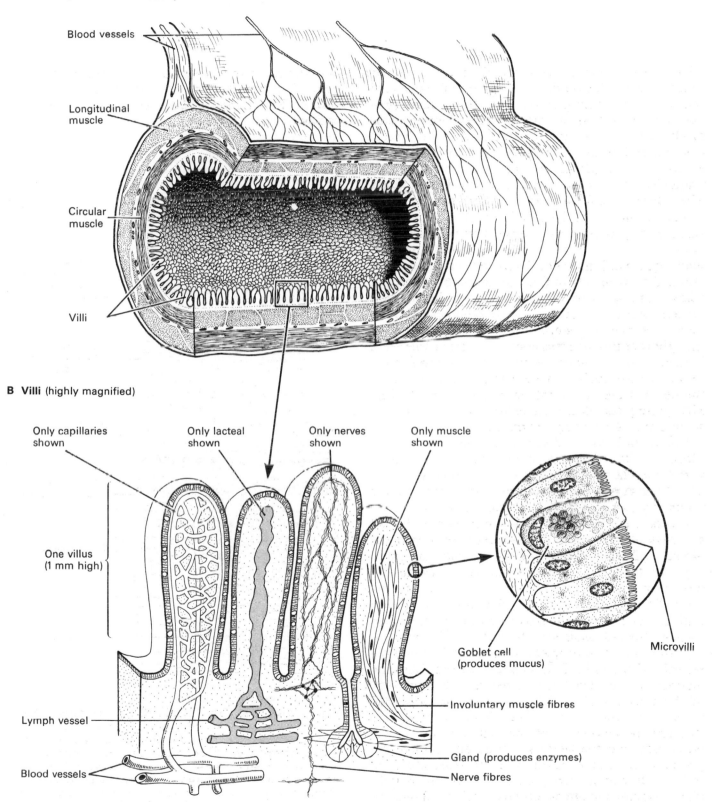

A A small piece of ileum cut open to show villi

Blood vessels

Longitudinal
muscle

Circular
muscle

Villi

B Villi (highly magnified)

Only capillaries
shown

Only lacteal
shown

Only nerves
shown

Only muscle
shown

One villus
(1 mm high)

Microvilli

Goblet cell
(produces mucus)

Lymph vessel

Involuntary muscle fibres

Blood vessels

Gland (produces enzymes)

Nerve fibres

Fig. 2.   Structure of the ileum wall.

The liver

The liver is the largest gland in the body. It weighs about 2 kg, is situated just below the diaphragm, and extends from one side of the body to the other (Fig. 26.3). The liver is dark red in colour owing to the large amount of blood it contains.

The liver makes and releases into the body many useful substances. The raw materials for these chemical processes come from food. The liver receives, via the **hepatic portal vein**, practically all the food absorbed by the intestine (Fig. 1).

The liver regulates the amount of food which reaches body cells. It does this by storing and sometimes processing the food it receives from the intestine, and then releasing it into the blood at a rate which depends upon the body's current needs. This and some other functions are described below.

Regulation of blood sugar

The liver, together with a set of glands in the pancreas, controls with great accuracy the amount of glucose sugar in the blood. It is very important that glucose is maintained at a certain constant level: first, because this sugar is the body's main source of energy; and second, because even slight changes in glucose concentration will alter the rate at which water moves in and out of cells by **osmosis**.

Whenever the blood's glucose level begins to rise, glands in the pancreas produce a substance called **insulin**. Insulin stimulates the liver cells to extract this glucose from the blood. At first the liver converts the glucose into glycogen (animal starch) and stores it in its cells. But the liver can hold only 100 g of glycogen. When this limit is reached any excess glucose remaining in the blood is converted into fat and transferred to more permanent storage areas under the skin and around various organs.

When glucose in the blood falls below the normal level, the pancreas slows down insulin production. The liver then converts its stored glycogen into glucose, which passes into the blood. When all the glycogen has been used up, stored fat is converted into glucose. If after prolonged starvation there is no more fat in the body, protein is converted into glucose. In this way the liver keeps the body supplied with food for as long as possible when food is not available elsewhere.

Regulation of amino acids and proteins

The body can store only very small amounts of amino acids and proteins. When a meal supplies more of these than the body can use the liver gets rid of them by a process called **deamination**.

Deamination is the removal from each amino acid molecule of the part which contains nitrogen: that is, the amino group (which has the chemical formula NH_2). These amino groups would automatically change into ammonia (NH_3) which is very poisonous, but the liver cells immediately convert them into urea, which is far less poisonous. Urea then passes from the liver into the blood and is eventually removed from the body by the kidneys. The remaining part of each amino acid molecule, which contains no nitrogen, is either converted by the liver into glucose and respired, or stored as glycogen.

Storage of vitamins and minerals

The liver stores vitamins A, D, and B_{12}, together with minerals such as iron, copper, and potassium until they are required by the body.

Purification of the blood

Many poisonous substances are produced by metabolism and by disease-causing organisms. Other poisons, including certain drugs and alcohol, are deliberately taken into the body. The liver can make most poisons harmless, after which they can be removed from the body by the kidneys.

Production of fibrinogen

The liver manufactures an important blood protein called **fibrinogen** which is vital to the clotting of blood in wounds. In this way the liver indirectly helps to preserve the body's supplies of blood and tissue fluid. Clotting of blood is described in Unit 36.

Production of heat energy

The liver produces a great deal of heat as a by-product of the thousands of chemical reactions which take place in its cells. This heat warms the blood as it passes through the liver, and the blood in turn warms tissues as it circulates around the body.

Production of bile

The liver excretes waste substances called bile pigments which are produced during the breakdown of old 'worn-out' red blood cells in the liver and spleen. Bile pigments are excreted into the intestine along with bile salts, also made by the liver, which emulsify fats and aid digestion (described in Unit 26).

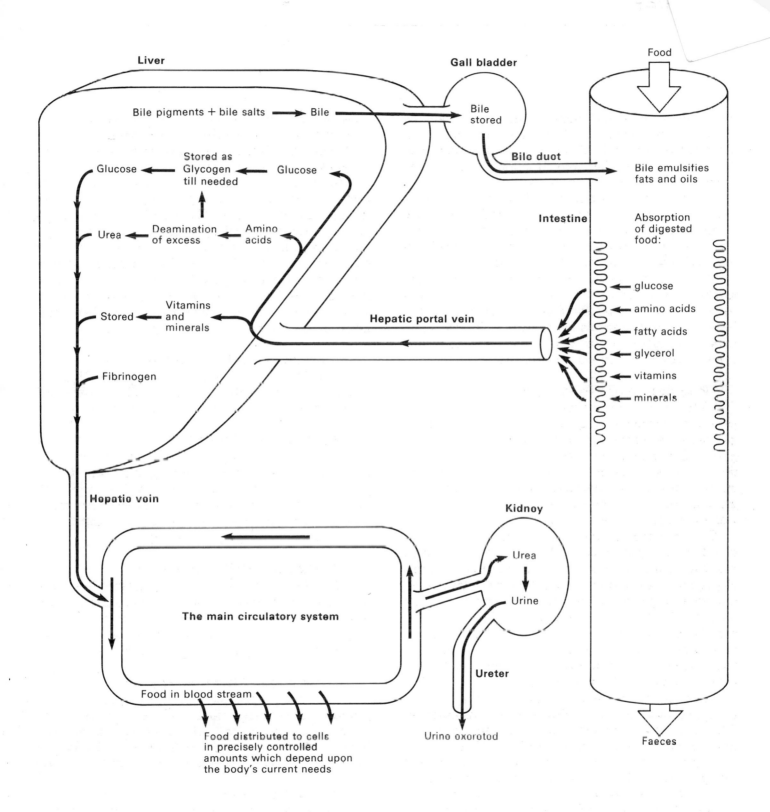

Fig. 1. The liver is the largest gland in the body. It receives, through the hepatic portal vein, almost all the food absorbed in the intestine, and then it releases it to the body at a rate which depends upon current needs. The liver keeps the amount of sugar in the blood at a fairly constant level by converting excess into glycogen or fat. It removes excess amino acids from the body by converting them into urea and glycogen. It stores vitamins and minerals, purifies the blood by neutralizing poisons, produces a great deal of heat, and makes fibrinogen, which is essential to the clotting of blood in wounds.

57

Blood and its functions

Adults have about 5·5 litres of blood. The blood's function is to transport substances around the body.

About 55 per cent of blood is a straw-coloured liquid called **plasma**. The remaining 45 per cent consists of red and white cells, sometimes called corpuscles, and tiny particles called **platelets**.

Plasma

Plasma is made up of water and many important dissolved substances. Plasma transports dissolved food from the intestine to all parts of the body, and carbon dioxide from all parts of the body to the lungs. It carries urea from the liver to the kidneys where it is excreted, and distributes chemicals called hormones from the endocrine glands to various tissues. Plasma also contains the protein fibrinogen which plays an important part in the clotting of blood in wounds.

Red blood cells, or erythrocytes

Human red cells are bi-concave discs (Fig. 1). There are about 5 million in a cubic millimetre of blood. They have no nucleus and live for only four months when they are destroyed in the spleen and liver. Some of their chemicals, such as iron, are re-used in the bone marrow for the manufacture of new red cells. About two million red cells are destroyed and replaced every second in the human body.

These cells owe their red colour to **haemoglobin**, a chemical which enables them to transport oxygen from the lungs to all parts of the body.

White blood cells, or leucocytes

White cells are really colourless, and unlike red cells, they have a nucleus. Most types of white cell can change shape, like an amoeba.

There are only 8000 white cells in a cubic millimetre of blood. About 75 per cent of white cells are made in the bone marrow; these cells have a large lobed nucleus, and are called **granulocytes** because they have tiny granules in their cytoplasm. The remaining 25 per cent of white cells are called **agranulocytes** because they have no granules. Most agranulocytes are made in the lymphatic tissues; therefore they are sometimes called **lymphocytes** (Fig. 2).

The majority of leucocytes are **phagocytic**, a word which means 'cell eater'. They are described in this way because they destroy invading bacteria and dead cells by engulfing and digesting them. **Neutrophils** and **monocytes** are examples of phagocytic white cells (Fig. 2). These cells gather in wounds where they kill bacteria before they can enter the body. They are also capable of squeezing between the cells which make up blood capillary walls in order to reach bacteria anywhere in the body (Fig. 3).

Platelets

Platelets are not complete cells. They are tiny fragments of cells made in the bone marrow. Their function during blood clotting is described in Unit 36.

A Front and side views of a red cell

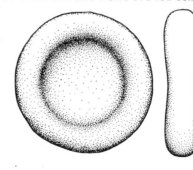

B Red cell cut in half

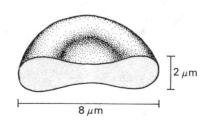

2 µm

8 µm

C Red cells as they appear in a blood clot

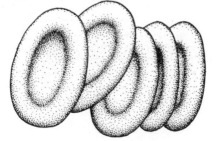

Fig. 1. Red blood cells have no nucleus and live for only four months, when they are destroyed in the spleen. New red cells are made in the bone marrow. Red cells contain haemoglobin, a chemical which enables them to transport oxygen from the lungs to all parts of the body.

Neutrophil (or polymorph) 60–70% of all white cells. Kill bacteria in wounds, in the bloodstream, and in tissue fluid

Eosinophil (or acidophil) 2% of white cells. Combat allergies, and increase in numbers during parasitic worm infections

Granulocytes
(lobed nucleus, granular cytoplasm, made in the bone marrow)

Basophil 0·5–2% of white cells. Probably make heparin (prevents blood clotting)

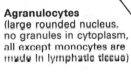

Monocyte 2–8% of white cells. Kill bacteria. Partly made in the bone marrow

Large and small lymphocytes
20–30% of white cells. Make antibodies

Agranulocytes
(large rounded nucleus, no granules in cytoplasm, all except monocytes are made in lymphatic tissue)

Fig. 2. The main types of white blood cell, or leucocyte, and their functions.

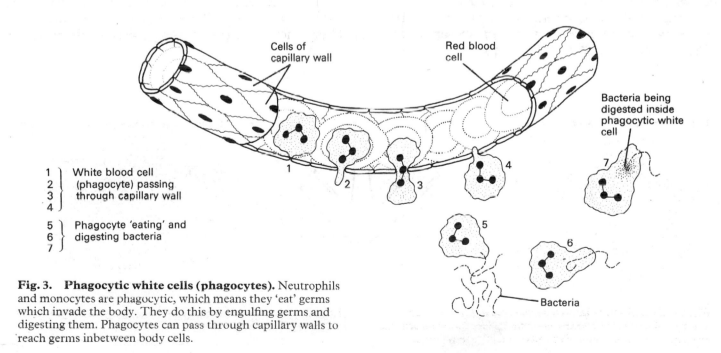

Cells of capillary wall

Red blood cell

Bacteria being digested inside phagocytic white cell

1 ⎫
2 ⎬ White blood cell
3 ⎪ (phagocyte) passing
4 ⎭ through capillary wall

5 ⎫ Phagocyte 'eating' and
6 ⎬ digesting bacteria
7 ⎭

Bacteria

Fig. 3. Phagocytic white cells (phagocytes). Neutrophils and monocytes are phagocytic, which means they 'eat' germs which invade the body. They do this by engulfing germs and digesting them. Phagocytes can pass through capillary walls to reach germs inbetween body cells.

Structure of the heart

The heart is a hollow bag with walls made of **cardiac muscle**. This type of muscle is described in Unit 8. When the cardiac muscle contracts it squeezes blood out of the heart into vessels called **arteries** which carry it to all parts of the body. When the cardiac muscle relaxes it fills with blood from vessels called **veins**. These return blood from its journey around the body. One contraction and relaxation of the heart is called a **heart beat**. At rest the heart beats 60 to 70 times a minute. During this time it pumps about 5 litres of blood. Exercise may increase this rate to about 150 times a minute and then blood flow increases to 20 litres a minute. The space inside the heart is divided into four compartments called **chambers**. The two top chambers are called **atria**, or **auricles**, and their walls contain a thin layer of cardiac muscle. The two lower chambers are called **ventricles** and their walls are made of a much thicker layer of cardiac muscle.

The heart contains a number of **valves**. These control the direction in which blood flows through the heart. The **bicuspid** and **tricuspid** valves ensure that blood flows only from the atria into the ventricles. The lower edges of these valves are held in place by tendons and muscles. **Semi-lunar** valves are situated at the points where blood flows out of the heart. The semi-lunar valves ensure that blood cannot flow back into the ventricles.

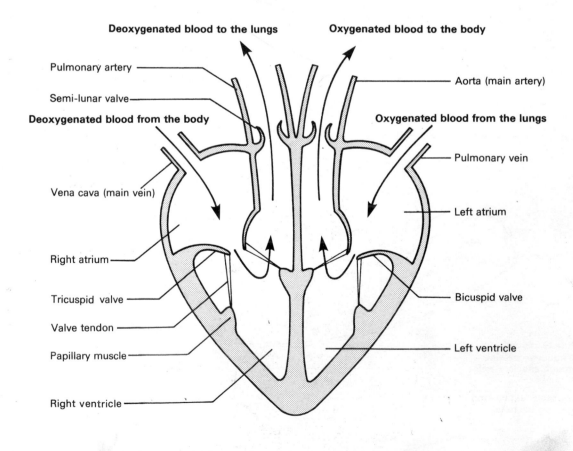

Deoxygenated blood to the lungs

Oxygenated blood to the body

Pulmonary artery

Semi-lunar valve

Deoxygenated blood from the body

Vena cava (main vein)

Right atrium

Tricuspid valve

Valve tendon

Papillary muscle

Right ventricle

Aorta (main artery)

Oxygenated blood from the lungs

Pulmonary vein

Left atrium

Bicuspid valve

Left ventricle

Fig. 1. Diagram of the heart showing the direction of the blood flowing through it. The heart is a bag whose walls are made of cardiac muscle. When this muscle contracts, blood is squeezed out of the heart and around the body. When the muscle relaxes, the heart fills with blood just returned from the body.

The heart has four compartments called chambers. Valves ensure that blood can only flow from the upper chambers—called atria—into the lower chambers—called ventricles. Valves at the exits from the heart ensure that blood cannot re-enter the heart by that route once it has left.

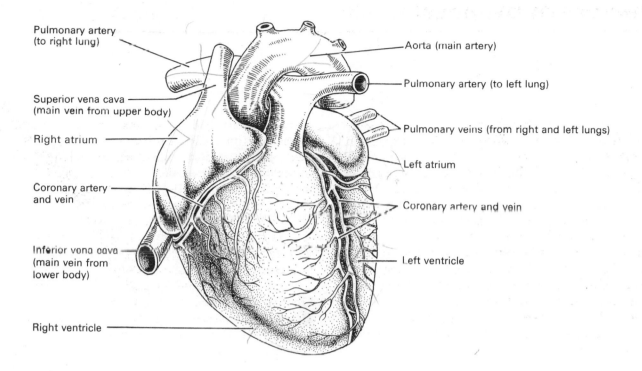

Fig. 2. **The human heart.** Note the coronary arteries and veins. These supply the heart's cardiac muscle with food and oxygen and remove its waste products.

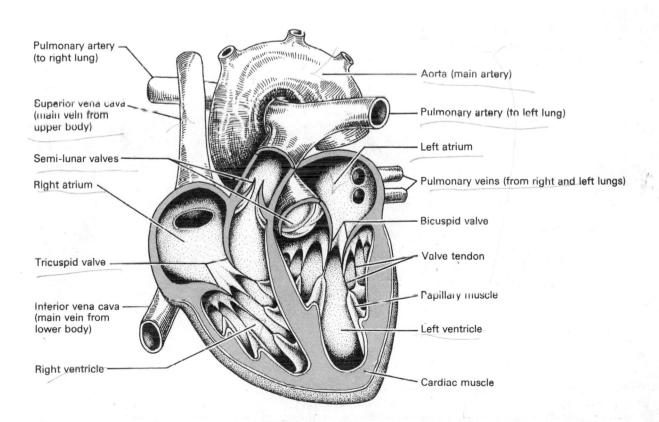

Fig. 3. **Drawing of a heart dissected to show its internal structure.**

How the heart pumps blood

A heart weighs only 300 g, and in the average lifetime it beats 2 500 000 000 times and pumps 340 million litres of blood.

The technical name for contraction of the heart is **systole**. This is followed by relaxation of the heart which is called **diastole**. One systole and diastole form what is known as a **cardiac cycle**. A cardiac cycle takes only 0·8 seconds to complete and during this time the following events occur.

Relaxation of the heart (diastole)

The atria and ventricles relax and increase in volume. At the same time the bicuspid and tricuspid valves open and blood flows into the heart from the veins (Fig. 1A).

Contraction of the heart (systole)

Both atria contract together forcing blood down into the ventricles. Simultaneously a ring of muscle contracts around each vein where it joins an atrium. This closes the veins and prevents blood from leaving the heart (Figs. 1B and 4B). A fraction of a second later the ventricles contract. They contract from the bottom upwards which forces blood up into the arteries past the semi-lunar valves (Fig. 1C). At the same time the bicuspid and tricuspid valves close, stopping blood entering the atria (Fig. 2). The artery walls are stretched outwards (distended) by the sudden increase in blood pressure (Fig. 4A).

The pulse

As the atria and ventricles relax, blood pressure in the heart falls below that in the arteries and the semi-lunar valves fill with blood. This prevents blood from flowing back into the heart (Figs. 3 and 4B). Elastic tissue in the artery walls was stretched when the arteries filled with blood and now it recoils (like someone letting go of a stretched elastic band), pressing inwards on the blood. Since the blood cannot return to the heart, it is forced out along the arteries (Fig. 4B). This sudden expansion of the arteries followed by elastic recoil causes a wave, or ripple, to move along their walls away from the heart. This ripple can be felt with the fingers in arteries at the wrist and neck, and is called the **pulse**.

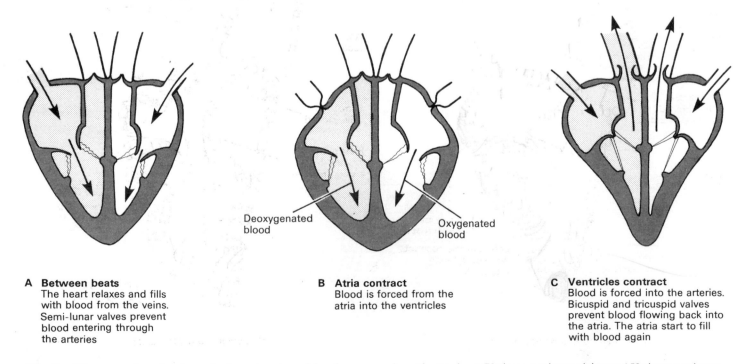

A Between beats
The heart relaxes and fills with blood from the veins. Semi-lunar valves prevent blood entering through the arteries

B Atria contract
Blood is forced from the atria into the ventricles

Deoxygenated blood

Oxygenated blood

C Ventricles contract
Blood is forced into the arteries. Bicuspid and tricuspid valves prevent blood flowing back into the atria. The atria start to fill with blood again

Fig. 1. Diagram showing how the heart pumps blood.
Contraction of the heart is called systole (**B** and **C** above). Relaxation between beats is called diastole (**A** above). At rest the heart beats about 70 times a minute rising to 150 times a minute during exercise.

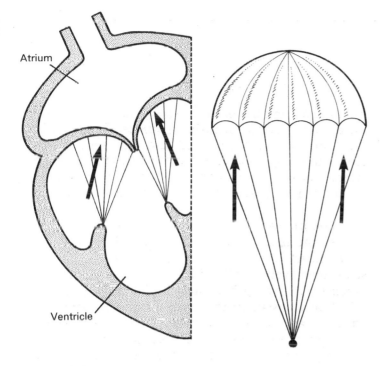

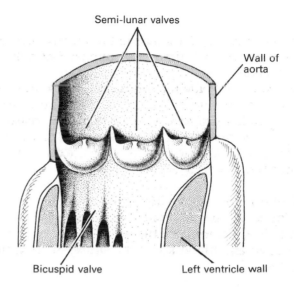

Fig. 2. Bicuspid and tricuspid valves can be compared with a parachute. During each heart beat, the valve flaps are pushed up to the limit of their tendons like a parachute filling with air.

Fig. 3. The main artery (aorta) cut open to show the semi-lunar valves. These are three pockets which fill with blood as the heart relaxes (at diastole) and close the aorta, preventing blood flowing back to the heart.

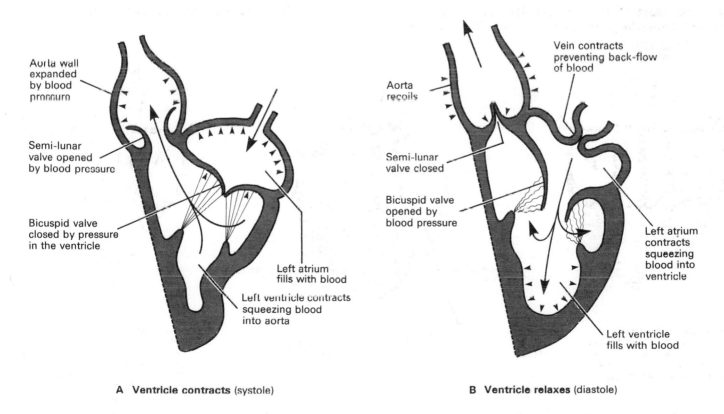

A Ventricle contracts (systole)

B Ventricle relaxes (diastole)

Fig. 4. Origin of the pulse. Artery walls expand when blood is forced into them (**A**). Then the stretched elastic walls recoil inwards squirting the blood along the artery (**B**). This sudden expansion and elastic recoil of an artery cause a ripple to travel along its walls away from the heart. This ripple causes the pulse, which can be felt at the wrist and neck.

Arteries, capillaries, and veins

The heart pumps blood into vessels called **arteries**. Arteries divide and sub-divide into narrower vessels called **arterioles** which, in turn, divide into extremely narrow vessels called **capillaries**. Eventually capillaries join together forming small veins called **venules**, and these lead into larger and larger veins which carry blood back to the heart.

Arteries and arterioles

Artery walls consist of three layers: an outer layer of fibrous tissue (elastic and collagen fibres); a middle layer of involuntary muscle and elastic fibres; and an inner layer only one cell thick called the **endothelium** which forms the lining of the vessel (Fig. 1A).

Walls of the larger arteries contain more elastic fibres than muscle. This allows them to expand and recoil with each heart beat as described in Unit 31. Small arteries and arterioles have more muscle than fibres in their walls. This muscle controls the diameter of the vessels making them narrower to reduce blood flow and wider to increase it.

Capillaries

Capillaries are the narrowest blood vessels (Fig. 1B). Some are so narrow that red blood cells are bent out of shape as they pass through. Capillary walls are a single layer of cells and are continuous with the lining of the larger vessels to which they are connected. Capillary walls are so thin that liquid from the blood is forced through them and then circulates between surrounding cells, carrying dissolved food and oxygen to them from the blood. This liquid is called **tissue fluid** and is described in Unit 34. Capillaries extend into almost every part of the body. None of the body's millions of cells is more than a fraction of a millimetre from one of these tiny vessels.

The amount of blood entering the capillaries is controlled by the diameter of the arterioles and by tiny rings of muscle where some arterioles join a capillary (Fig. 3). These rings of muscle are called **precapillary sphincters**. Flow of blood into a capillary network can be slowed down or stopped by these sphincters, and the blood diverted to other parts of the body.

Veins

Veins have thinner walls and wider passageways than arteries (Fig. 1C). Blood in veins is at a much lower pressure than in arteries, because pressure is lost as blood passes through the vast network of capillaries on its way to the veins. Flow of blood along veins to the heart is assisted in three ways. First, veins contain pocket valves which allow blood to flow only towards the heart. Second, many large veins are situated inside muscles of the legs, thighs, arms, etc. When these muscles are flexed the veins are squeezed and this squirts blood towards the heart. The valves prevent it flowing in the opposite direction (Fig. 2). If it were not for this mechanism, blood would gather in the feet and legs. Muscles are never still for long, even during sleep, and so blood flows along the veins at all times. Third, as the heart relaxes (diastole) its blood pressure is lower than in the veins and this causes blood to be sucked from the vein into the heart.

If the vein valves fail to close completely, blood gathers in the veins causing them to swell and become painful. This condition is called **varicose veins**. It is most likely to occur in the leg veins of people who have to spend long periods standing, or who frequently lift heavy weights. Such people should exercise their legs regularly, avoid standing still in one position for a long time, and rest with their feet up whenever possible.

A Structure of an artery
The wall consists of three layers

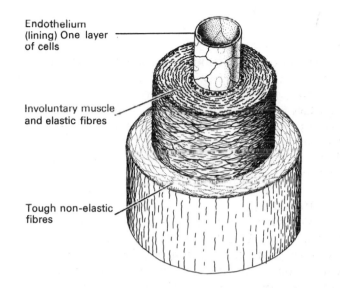

Endothelium
(lining) One layer
of cells

Involuntary muscle
and elastic fibres

Tough non-elastic
fibres

B A capillary (not drawn to scale)
Wall consists of one layer of cells

C Structure of a vein
The two outer layers are thinner
than those in an artery wall

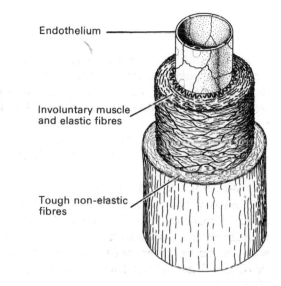

Endothelium

Involuntary muscle
and elastic fibres

Tough non-elastic
fibres

Fig. 1. Arteries, capillaries and veins. Arteries have thick walls and carry blood at high pressure away from the heart. Arteries divide to form extremely narrow vessels called capillaries. Capillary walls are so thin that liquid passes through them from the bloodstream, carrying dissolved food and oxygen to surrounding cells. Capillaries join up to form veins. Veins carry blood at low pressure back to the heart and they contain valves which ensure that blood flows in only that direction.

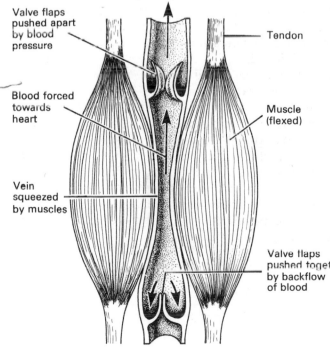

Valve flaps
pushed apart
by blood
pressure

Blood forced
towards
heart

Vein
squeezed
by muscles

Tendon

Muscle
(flexed)

Valve flaps
pushed together
by backflow
of blood

Fig. 2. Many large veins are situated between muscles. When the muscles are flexed the veins are squeezed, forcing blood in the direction which the valves allow—that is, towards the heart.

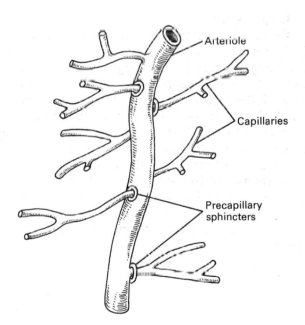

Arteriole

Capillaries

Precapillary
sphincters

Fig. 3. Control of blood flow through capillaries. The amount of blood entering capillaries is controlled by the diameter of the arterioles and by tiny rings of muscle called precapillary sphincters where some capillaries branch off from arterioles. These sphincters can slow down or stop flow of blood into a capillary network, thus diverting blood to other parts of the body.

The main blood vessels

Humans and all other mammals have a *double* circulatory system: their blood circulates through two separate systems joined only at the heart. One system is called the **pulmonary circulation** and carries blood from the heart to the lungs and back. The other system is called the **systemic circulation** and carries blood from the heart to all parts of the body except the lungs and back again (Fig. 1).

Pulmonary circulation

Starting at the right ventricle, blood is pumped at high pressure along the pulmonary artery to the lungs. This blood is **deoxygenated**, which means it contains no oxygen, having released all of this gas on its journey around the body. The blood renews its supply of oxygen as it passes through the lungs; that is, it becomes **oxygenated** again. It then returns at low pressure to the heart through the pulmonary vein.

Systemic circulation

Blood from the lungs enters the heart at the left atrium. It is pumped to the left ventricle and then pumped at very high pressure into the systemic circulation through the **aorta**, or main artery of the body. The aorta has branches leading to the head and arms, to the legs, and to organs in the abdomen (Fig. 2).

Two large veins return blood to the heart and complete the systemic circulation. The **superior vena cava** is the vein which carries blood from the head and arms, and the **inferior vena cava** returns blood from the legs and abdomen. Both of these veins empty into the right atrium.

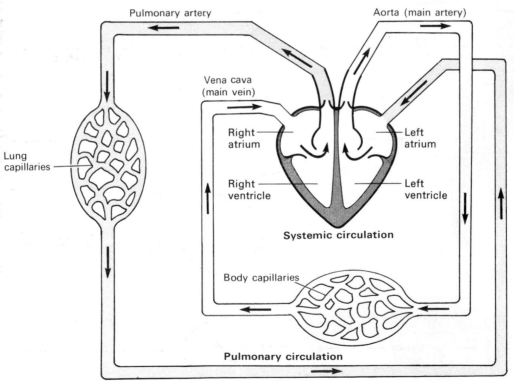

Fig. 1. Diagram of the double circulatory system of humans and other mammals. The pulmonary circulation carries blood from the heart to the lungs and back again. The systemic circulation carries blood from the heart to all parts of the body except the lungs and back again.

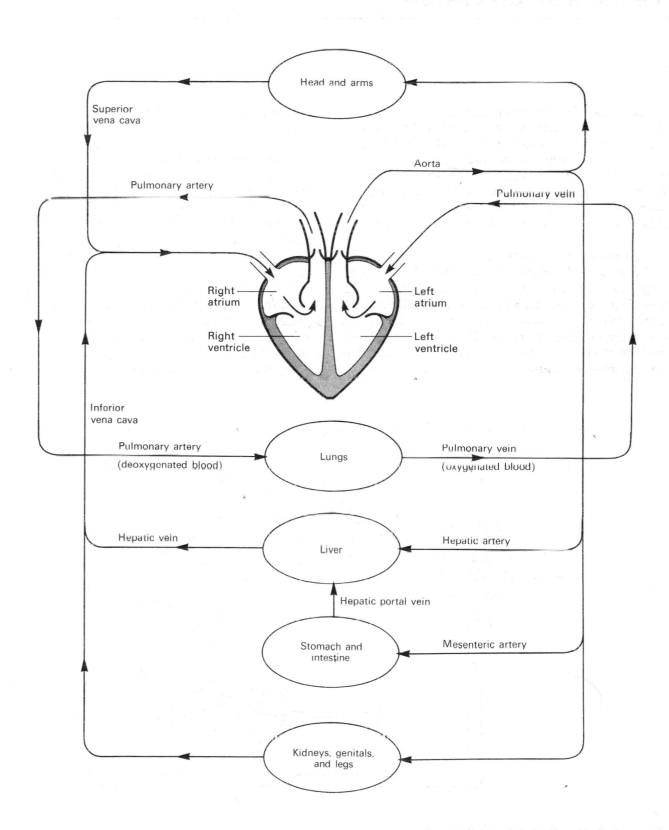

Fig. 2. Diagram of the main blood vessels.

Tissue fluid and lymph

Tissue fluid

Blood containing oxygen and food travels at high pressure through arteries to all parts of the body. Arteries divide into networks of capillaries known as **capillary beds** (Fig. 1). Blood pressure is so high at the artery end of a capillary bed and the capillary walls are so thin that a clear liquid is forced through them from the blood into the spaces between surrounding cells. This liquid is called **tissue fluid**.

As tissue fluid filters through the capillary walls it carries with it oxygen, food, and other useful substances from the bloodstream to the cells. As tissue fluid moves between the cells it distributes these useful substances, and at the same time it carries away carbon dioxide and other waste substances produced by the cells.

After picking up cell waste substances, some of the tissue fluid passes back into the bloodstream through capillary walls at the vein end of each capillary bed (Figs. 1 and 2). Here, blood pressure is very much lower than at the artery end of the capillary bed. The remaining tissue fluid drains into the lymphatic system and becomes lymph.

Lymph

The lymphatic system begins as narrow, thin-walled tubes called **lymph capillaries** (Figs. 1 and 2). These are almost as numerous in the tissues as blood capillaries. Lymph capillaries drain into larger lymph vessels similar to veins: they have connective tissue in their walls, they have valves, and they are situated between muscles. When the muscles contract they squeeze the lymph vessels and force lymph through them in the direction which the valves permit. The lymphatic system empties its contents into the bloodstream at the large veins in the neck (Fig. 3).

The main function of the lymphatic system is to return to the bloodstream tissue fluid containing substances which cannot re-enter it at the vein end of capillary beds. These substances are mainly protein molecules which can sometimes leak out through capillary walls but which are too large to return in the opposite direction. These molecules can, however, pass through lymph capillary walls and return to the blood through the lymphatic system.

Before lymph returns to the blood it passes through a number of **lymph nodes** (Figs. 2 and 3).

These contain passageways with special phagocytic white cells attached to their walls. These white cells engulf and destroy bacteria and any other solid particles in the lymph, thereby helping to prevent infection.

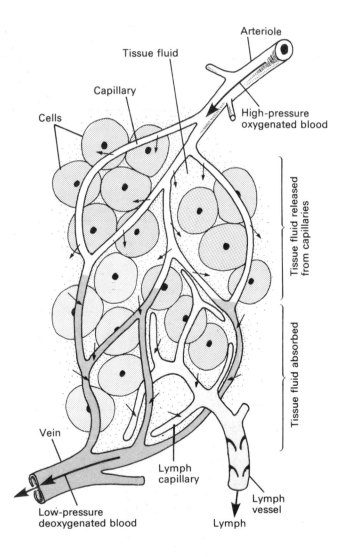

Fig. 1. Diagram of a capillary bed: a network of capillaries running between cells. High pressure at the artery end of the capillaries forces liquid called tissue fluid through capillary walls. Tissue fluid carries food and oxygen from the blood to the cells and collects carbon dioxide and other waste from the cells. Some tissue fluid is absorbed back into capillaries at the vein end of the bed, and the remainder is absorbed into lymph capillaries.

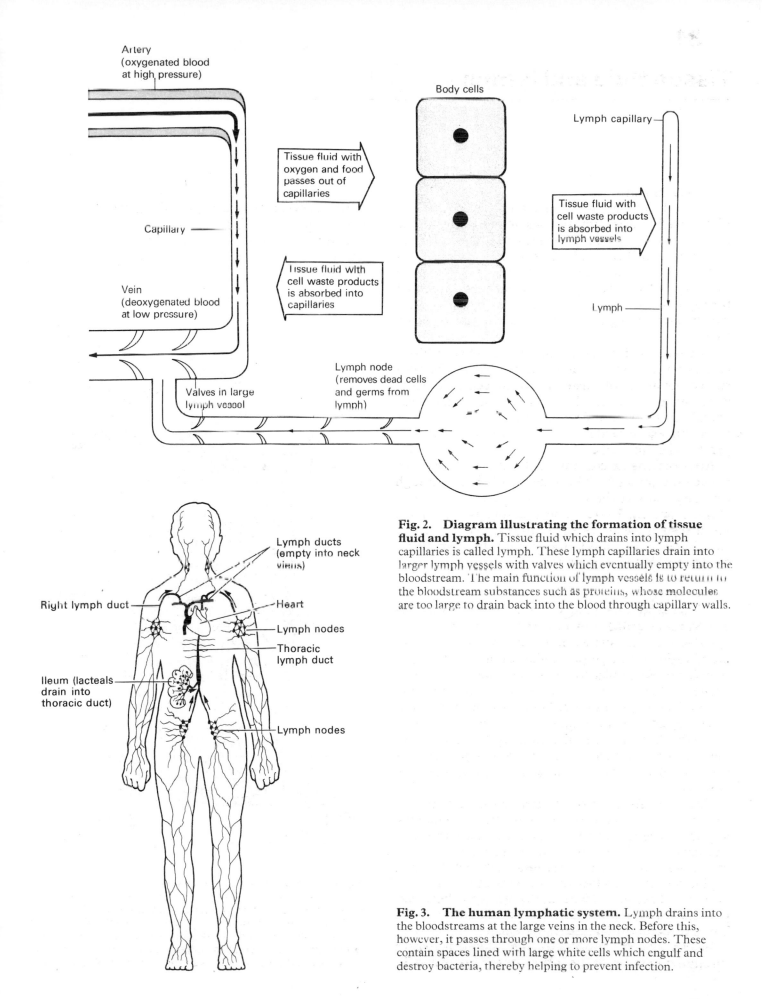

Artery
(oxygenated blood
at high pressure)

Body cells

Lymph capillary

Tissue fluid with
oxygen and food
passes out of
capillaries

Capillary

Tissue fluid with
cell waste products
is absorbed into
lymph vessels

Tissue fluid with
cell waste products
is absorbed into
capillaries

Vein
(deoxygenated blood
at low pressure)

Lymph

**Valves in large
lymph vessel**

Lymph node
(removes dead cells
and germs from
lymph)

Fig. 2. Diagram illustrating the formation of tissue fluid and lymph. Tissue fluid which drains into lymph capillaries is called lymph. These lymph capillaries drain into larger lymph vessels with valves which eventually empty into the bloodstream. The main function of lymph vessels is to return to the bloodstream substances such as proteins, whose molecules are too large to drain back into the blood through capillary walls.

Lymph ducts
(empty into neck
veins)

Right lymph duct

Heart

Lymph nodes

**Thoracic
lymph duct**

**Ileum (lacteals
drain into
thoracic duct)**

Lymph nodes

Fig. 3. The human lymphatic system. Lymph drains into the bloodstreams at the large veins in the neck. Before this, however, it passes through one or more lymph nodes. These contain spaces lined with large white cells which engulf and destroy bacteria, thereby helping to prevent infection.

69

Blood groups and blood transfusions

If more than 40 per cent of someone's blood is lost over a short period of time the victim's body cannot make new blood quickly enough to make up the loss before death occurs. But death from loss of blood can be prevented by a transfusion of blood from another person called a **blood donor**.

Blood from one person can only be transfused into certain other people. This is because there are four types of blood and when some types are mixed the red cells stick together in large clumps. This is called **agglutination**. Later the red cells split open spilling their contents into the plasma, and if this happened inside a patient's body during a blood transfusion it could seriously damage the kidneys.

Blood groups

The four types of blood are called **blood groups**, and are known by the letters A, B, AB, and O. Before a blood transfusion can take place it is necessary to make sure that the donor's and recipient's blood will mix together without agglutination. Blood groups which mix without agglutination are said to be **compatible**.

Blood compatibility depends upon chemicals called **antigens** on the surface of the red cells, and chemicals called **antibodies** in the plasma. There are two types of antigen: A and B; and two types of antibody: anti-A and anti-B.

Blood group A has A antigen on its red cells and anti-B antibody in its plasma.

Blood group B has B antigen on its red cells and anti-A antibody in its plasma.

Blood group AB has A and B antigens on its red cells and no antibodies in its plasma.

Blood group O has no antigens on its red cells and both anti-A and anti-B antibodies in its plasma.

Figure 1 shows how to discover your blood group.

Blood transfusions

Anti-A plasma agglutinates A red cells, and anti-B plasma agglutinates B red cells. So these combinations of plasma and red cell are incompatible as far as blood transfusion is concerned.

These facts have given rise to a rule for blood transfusions: the donor's red cells must be compatible with the recipient's plasma. Thus, a recipient with anti-A plasma can only receive blood with either B red cells, or red cells without antigens; and a recipient with anti-B plasma can only receive blood with either A red cells, or red cells without antigens. Blood can be safely transfused as follows:

Blood group	Can be transfused into	Can receive blood from
A	A and AB	A and O
B	B and AB	B and O
AB	AB only	All groups
O	All groups	O only

People with Group O blood are called **universal donors**. Their red cells have no antigens and so cannot be agglutinated by blood of any other group. People with Group AB are called **universal recipients**. Their plasma has no antibodies therefore it does not agglutinate blood from the other groups.

The Rhesus factor

About 85 per cent of humans have an antigen on their red cells called the Rhesus factor. People with the Rhesus factor are called **Rhesus positive** (Rh+) and those without it are called **Rhesus negative** (Rh−).

Rh− patients can receive one transfusion of Rh+ blood without harm because their plasma does not have an antibody to react with the incoming red cells. Subsequent transfusion, however, may be dangerous because Rh+ blood stimulates the body of a Rh− recipient into producing a plasma antibody which agglutinates Rh+ blood. Such recipients are then said to be **sensitized** to Rh+ blood and will agglutinate any which is transfused into them. Rh− blood can be transfused into Rh+ people any number of times without harm.

A Rh+ father and a Rh− mother could have a Rh+ child. During pregnancy the developing child's red cells may enter the mother's blood, perhaps through a fault in the placenta. The mother's body will then produce an antibody which destroys Rh+ cells. This antibody will not harm her first child but if she has a second Rh+ child and its red cells enter her blood she will produce more antibody and there is a danger that this will reach the embryo, destroy its red cells, and cause serious blood disorders or death.

This danger can now be avoided. A Rh− mother with a new born Rh+ child can be injected with chemicals which stop her body producing the Rhesus antibody.

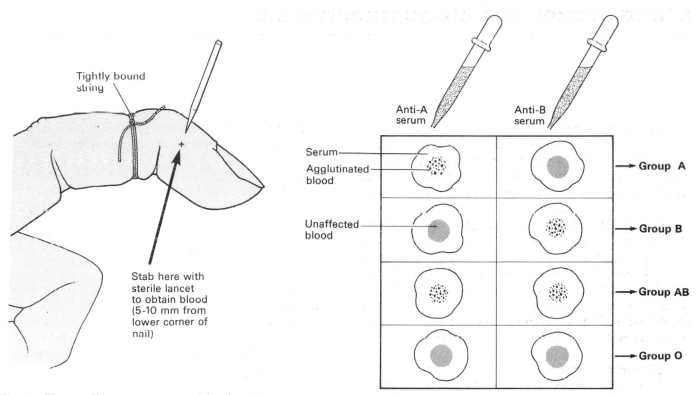

Fig. 1. How to discover your own blood group.

A Bind some string around a finger as shown below. Stab the area of skin indicated by the cross on the drawing with a sterile lancet to obtain blood.

B Put two drops of blood on a white tile. Place Anti-A serum on one drop of blood and Anti-B serum on the other. All possible results are shown above.

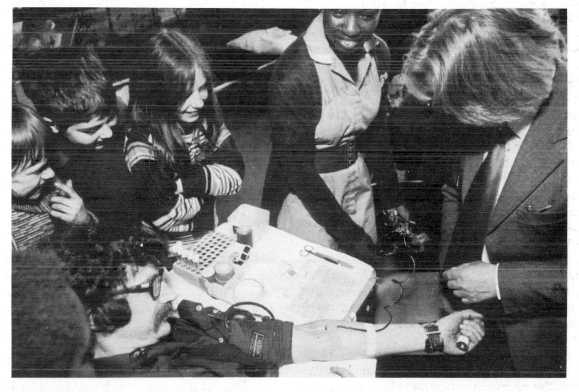

A blood donor donating blood.

Young children watch as a blood donor gives blood. Each donor gives about 420 cm³ of blood through a vein in the arm (notice the bag of blood which the nurse is holding). Sodium citrate is added to prevent clotting and it is stored at 5 °C in sterilized bottles. Within about a month a donor will have replaced the lost blood.

36

Blood clotting and treatment of bleeding and wounds

Blood clotting

Cuts bleed for a while and then the blood changes into a thick jelly called a **blood clot**. The clot closes the injured blood vessels preventing further loss of blood. Later the clot dries and hardens preventing germs and dirt from entering the wound while healing takes place.

A complicated series of events take place during the clotting of blood. Briefly this is what happens.

A soluble blood protein called **fibrinogen** is converted into a net of tangled threads made of an *insoluble* protein called **fibrin**. These sticky fibrin threads radiate out from the platelets like the spokes of a wheel (Fig. 2). They adhere to surrounding tissues and trap blood cells. A jelly-like mass of blood cells and fibres is formed, which dries into a blood clot.

Obviously it is vital that fibrin threads only form when vessels are damaged. At any other time they would clog up the circulatory system.

Clotting does not normally occur in undamaged vessels because fibrinogen can only be changed into fibrin by the enzyme called **thrombin**, and in undamaged vessels thrombin exists as an *inactive* substance called **prothrombin**.

When a wound occurs platelets and surrounding damaged tissues release the enzyme **thromboplastin**. In the presence of calcium ions, thromboplastin converts prothrombin to thrombin. Thrombin then converts fibrinogen into fibrin, and fibrin threads cause a blood clot to form (Fig. 1).

Recent research has shown that the formation of thromboplastin by platelets and damaged tissue results from complex reactions involving a number of **blood factors**.

Haemophilia and blood factor VIII

Haemophilia is a blood disorder which male children can inherit from their mothers. The blood of haemophiliacs clots very slowly because it lacks factor VIII (anti-haemophiliac globulin), so very little thromboplastin is formed when blood vessels are damaged.

Haemophiliacs may suffer extensive bleeding from wounds, but the biggest danger is bleeding inside damaged joints which, if not treated, can lead to permanent crippling.

Treatment involves supplying haemophiliacs with blood factor VIII, which can be obtained from fresh plasma.

Treatment of bleeding and wounds

This subject is dealt with very briefly in Fig. 3.

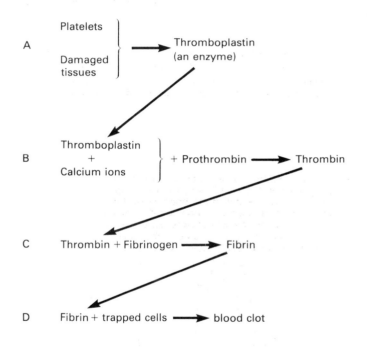

Fig. 1. **Diagram of the changes which cause blood to clot in wounds.** When blood vessels are damaged the soluble protein fibrinogen is converted into insoluble threads of fibrin. Fibrin threads make a sticky net which traps blood cells and forms a blood clot. Fibrinogen is changed into fibrin by the enzyme thrombin, but in undamaged vessels thrombin exists as an inactive substance called prothrombin. When a wound occurs platelets and damaged tissues release thromboplastin (A above), an enzyme which converts prothrombin to thrombin (B), which converts fibrinogen to fibrin (C), which causes blood to clot (D).

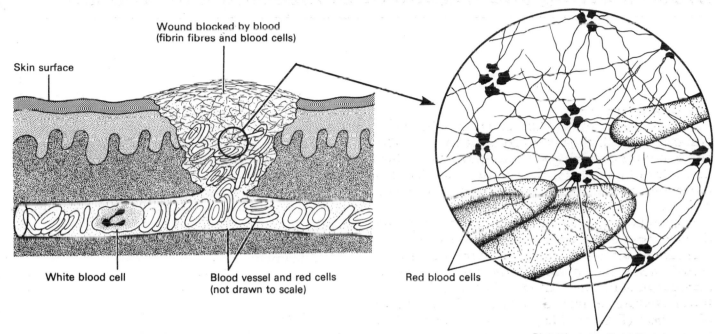

Wound blocked by blood
(fibrin fibres and blood cells)

Skin surface

White blood cell

Blood vessel and red cells
(not drawn to scale)

Red blood cells

Platelets (with fibrin fibres
radiating from them)

Fig. 2. A blood clot. Threads of fibrin radiate out from the platelets. These threads trap red cells, thus blocking up the wound and stopping further loss of blood. The clot dries out forming a scab which stops germs entering the wound while it heals.

A To stop bleeding Press a pad of clean cloth onto the wound. If a limb is badly cut it helps to lift it above the rest of the body. Keep the patient still and quiet. Call a doctor: if bleeding does not stop; if you suspect internal bleeding; if there is something embedded in the wound; if the wound has been caused by an animal

B Dressing a wound First wash your hands. Then clean the skin around the wound and the wound itself, using running water and then mild antiseptic. When dry, cover the wound with sterile gauze and a pad of cotton wool if the wound is large. Hold in place with adhesive tape or with a bandage (as shown in the drawings) but do not bind too tightly

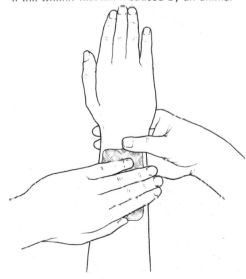

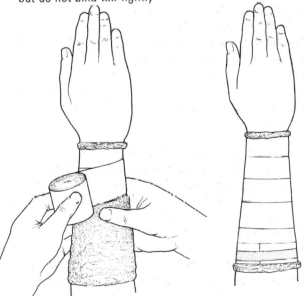

Fig. 3.   Treatment of bleeding and wounds.

37

Structure of the respiratory system

The respiratory system is made up of air passages through the nose, a tube called the wind-pipe, or **trachea**, and the lungs. The lungs are situated in a space inside the chest called the **thoracic cavity**. The walls of this space are supported by the ribs and its floor is a sheet of muscle called the **diaphragm** (Fig. 1).

The inner surface of the thoracic cavity and the outer surface of the lungs are covered with a shiny slippery skin called the **pleural membrane** (Fig. 1). It produces **pleural fluid**. This fluid acts as a lubricant, reducing friction as the lungs rub against the ribs during breathing.

Air breathed in through the nose is warmed, made moist, and has dust and germs removed from it before it reaches the lungs. The air passages down to the lungs are lined with a carpet of microscopic hairs called **cilia** (Fig. 2C). Between the cilia are cells which produce sticky liquid called **mucus** in which germs and dust are trapped. Cilia wave back and forth like oars, moving the mucus with trapped germs and dust up to the throat, where it is swallowed.

The wind-pipe has walls supported by curved strips of cartilage (Fig. 1). These hold the wind-pipe open even when the neck is bent. The voice box, or **larynx**, is at the top of the trachea. Membranes called **vocal cords** are attached to the sides of the voice box. Normally these membranes are slack and air passes soundlessly between them. But when muscles pull them taut and air is forced between them, they vibrate and make sounds which form the voice.

At its base the wind-pipe divides into two tubes called **bronchi** (Figs. 1 and 2). One bronchus leads into each lung. Inside the lungs the bronchi divide like the branches of a tree into extremely narrow tubes called **bronchioles**. The smallest bronchioles have muscles in their walls which can make them wider or narrower, thereby altering the rate of air flow in and out of the lungs.

Bronchioles end in tiny bubble-like air sacs called **alveoli** (Fig. 2B). Each alveolus is about 0·2 mm in diameter and there are about 300 million in a set of human lungs. Alveoli give the lungs a spongy texture. A network of capillaries surrounds each alveolus (Fig. 2B). The blood in these capillaries absorbs oxygen from air breathed into the alveoli, and releases carbon dioxide into the air breathed out of the alveoli.

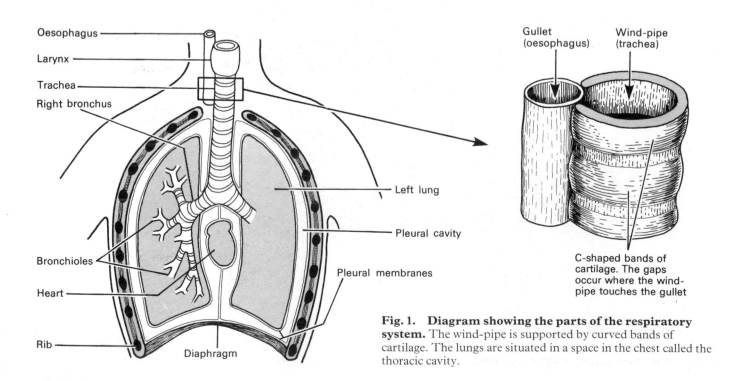

Fig. 1. Diagram showing the parts of the respiratory system. The wind-pipe is supported by curved bands of cartilage. The lungs are situated in a space in the chest called the thoracic cavity.

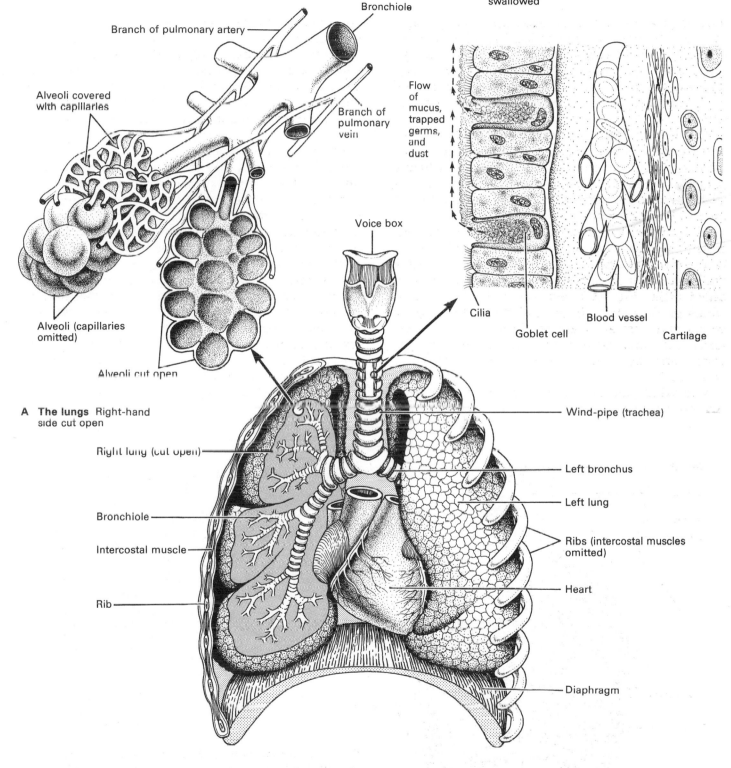

B Bronchioles and alveoli Bronchioles end in air sacs called alveoli. Each alveolus is covered with capillaries

Branch of pulmonary artery

Bronchiole

Alveoli covered with capillaries

Branch of pulmonary vein

Alveoli (capillaries omitted)

Alveoli cut open

A The lungs Right-hand side cut open

Right lung (cut open)

Bronchiole

Intercostal muscle

Rib

Voice box

C Wall of wind-pipe (highly magnified) It is lined with cilia and goblet cells, which produce mucus. Cilia carry the mucus with trapped dust and germs to the back of the mouth where they are swallowed

Flow of mucus, trapped germs, and dust

Cilia

Goblet cell

Blood vessel

Cartilage

Wind-pipe (trachea)

Left bronchus

Left lung

Ribs (intercostal muscles omitted)

Heart

Diaphragm

Fig. 2.   Structure of the respiratory system.

75

Air is moved in and out of the lungs by the action of the **diaphragm** and **intercostal muscles**.

Inspiration (breathing in)

The diaphragm is dome-shaped and relaxed before a breath is taken (Fig. 1B). During the first stage of breathing in, the diaphragm **contracts** and becomes flatter in shape (Fig. 1A). At the same time **external intercostal muscles** (Fig. 4) contract, pulling the ribs upwards so that they pivot where they join the backbone (Fig. 3). The flattening of the diaphragm and the lifting of the rib cage increase the volume of the thoracic cavity. This is automatically followed by an increase in lung volume. The increase in volume reduces air pressure inside the lungs, and air rushes into them from the atmosphere through the air passages.

Expiration (breathing out)

The diaphragm and external intercostal muscles relax. The rib cage drops and the diaphragm returns to its original dome shape. These movements reduce the volume of the thoracic cavity and the lungs return to their original volume. This squeezes air out of the lungs through the air passages. Air can be forced out of the lungs by contracting the **internal intercostal muscles**. This happens, for example, when playing a wind instrument.

At rest, an adult breathes about 16–18 times a minutes, using the diaphragm alone. During exercise this rate increases, and the intercostal muscles are used to increase the volume of each breath. This supplies the muscles with extra oxygen and removes carbon dioxide more quickly.

Figure 2 demonstrates how the intercostal muscles work. Fit an elastic band through holes A and B and watch how the length changes as you move the apparatus. The elastic band represents the external intercostal muscles (which raise the rib cage). Then fit the elastic band through holes C and D. It now represents the internal intercostal muscles (which lower the rib cage, forcing air from the lungs).

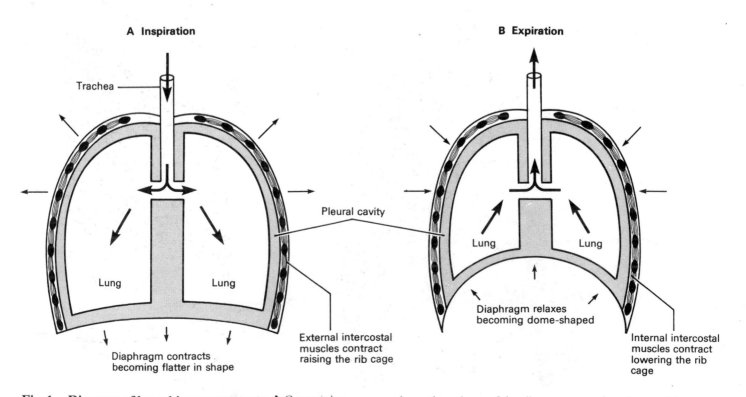

A Inspiration

Trachea

Pleural cavity

Lung Lung

Lung Lung

Diaphragm contracts becoming flatter in shape

External intercostal muscles contract raising the rib cage

B Expiration

Diaphragm relaxes becoming dome-shaped

Internal intercostal muscles contract lowering the rib cage

Fig. 1. Diagram of breathing movements. A Contraction of the intercostal muscles and diaphragm increases the volume of the rib cage inflating the lungs. **B** Relaxation of these muscles reduces the volume of the rib cage, squeezing air out of the lungs.

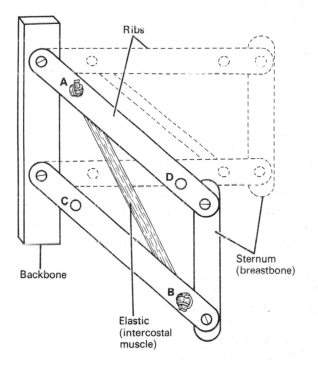

Fig. 2. Model showing how intercostal muscles move the rib cage.

Labels in Fig. 2: Ribs; A; D; C; B; Backbone; Sternum (breastbone); Elastic (intercostal muscle)

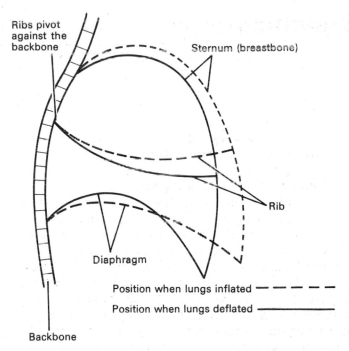
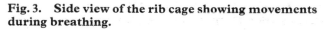

Fig. 3. Side view of the rib cage showing movements during breathing.

Labels in Fig. 3: Ribs pivot against the backbone; Sternum (breastbone); Rib; Diaphragm; Backbone; Position when lungs inflated -----; Position when lungs deflated ———

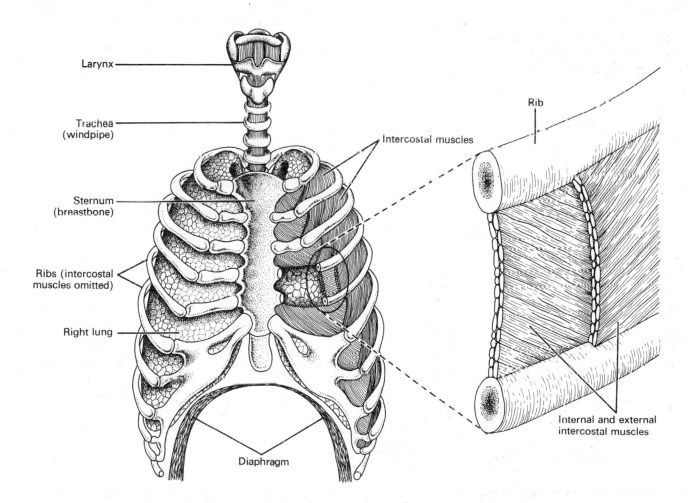

Fig. 4. Rib cage, showing the position of intercostal muscles.

Labels in Fig. 4: Larynx; Trachea (windpipe); Sternum (breastbone); Ribs (intercostal muscles omitted); Right lung; Diaphragm; Intercostal muscles; Rib; Internal and external intercostal muscles

77

39

Lung volume, breathing rate, and artificial respiration

Lung volume

The total average volume of human lungs is 4·5 litres in females and 5·5 litres in males. But in normal breathing only 500 cm³, or half a litre, of air is breathed in and out. This is called **tidal air**.

The greatest amount of air which can be inhaled into and forced out of the lungs is called the **vital capacity** of the lungs (Fig. 1). Vital capacity varies between 4 and 5 litres, depending upon age and fitness. It is greatly increased by regular strenuous exercise.

No matter how hard you try it is impossible to force all the air from your lungs. The remaining 1 to 1·5 litres which cannot be exhaled is called **residual air**. Residual air is not stagnant, it is mixed with inhaled air at each breath.

Regulation of breathing rate

An adult at rest breathes about 16 to 18 times a minute. This rate is increased by exercise, disease, and shocks such as pain or a sudden chill. Breathing can be increased or decreased at will but normally it is controlled by two unconscious mechanisms which work together.

Nervous control of breathing Breathing rate is controlled by the part of the brain called the **respiratory centre**. This centre controls the diaphragm and intercostal muscles and it is connected to nerves called **stretch receptors** in the walls of the bronchioles and alveoli. As the lungs fill with air, these receptors are stretched and send off many nerve impulses to the respiratory centre. When the lungs are deflated the receptors are relaxed and send out few impulses. The number of impulses from the stretch receptors provides the respiratory centre with information about the amount of air in the lungs. It uses this information to increase or decrease the rate of breathing and volume of each breath according to the body's oxygen needs.

Chemical control of breathing The respiratory centre is sensitive to changes in the amount of carbon dioxide present in the blood. During exercise there is a rise in the level of blood carbon dioxide owing to increased respiration in the muscles. The respiratory centre detects this rise and automatically increases the rate and depth of breathing. This removes the excess carbon dioxide from the body and at the same time provides it with extra oxygen for respiration in the muscles.

Certain arteries in the neck contain nerve cells sensitive to changes in the amount of oxygen in the blood. At high altitudes, for instance, low air pressure can greatly reduce the rate at which oxygen is absorbed into the blood. This is detected by the nerve cells which send impulses to the respiratory centre, resulting in an increased rate of breathing.

Artificial respiration

It is often possible to restore normal breathing in someone rescued from drowning, for example, or who has received an electric shock, by various methods of artificial respiration. One of the most successful methods is called mouth-to-mouth resuscitation.

1. Make sure the victim's mouth and throat are clear of obstructions such as false teeth or pond weed.

2. Lay the victim face upwards and hold the head well back (Fig. 2 A and B). This ensures the wind-pipe is kept open and avoids the tongue falling backwards and blocking the throat (as often happens when an unconscious person is laid on his back).

3. Pinch the victim's nose so that no air can escape through it, and form an airtight seal with your mouth around the victim's mouth. Blow air gently into the mouth. This will inflate the lungs and you will see the chest rise (Fig. 2 C and D).

4. Repeat about 12 times a minute with adults and 20 times a minute with children until normal breathing begins.

Mountain sickness

Air pressure, and the amount of oxygen available, are reduced as a climber ascends a mountain or as an aeroplane gains altitude. Above 3000 m lack of oxygen makes clear thinking difficult and can cause giddiness or sickness, and above 6000 m it can cause permanent damage to the brain.

But the body can adapt to these conditions. Mountain people who live most of their lives above 3000 m have larger lungs, breathe more deeply, and have more red blood cells and a greater output of blood from the heart than people who live at sea level.

The photograph and caption opposite explain what can happen when deep-sea divers experience very high air pressures.

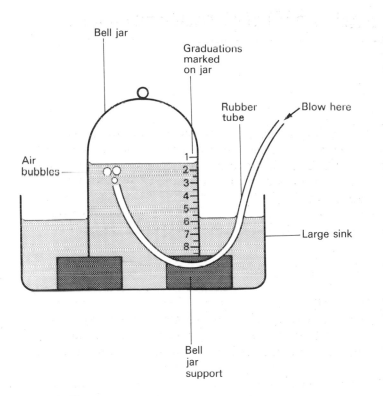

Fig. 1. How to measure lung vital capacity. Mark 1 litre graduations on a bell jar by pouring in one litre of water at a time. When the bell jar is full, stand it on supports (e.g. metal blocks) in a large sink of water. Take a deep breath, then exhale fully through a rubber tube leading into the bell jar. Estimate your lung vital capacity by means of the 1 litre graduations.

A deep-sea diver breathes air at up to four times normal pressure. The deeper he dives, the greater the pressure. This increased pressure causes the gases he breathes to dissolve in his blood. At depths of 50 m or more, nitrogen becomes poisonous, so deep-sea divers breathe a mixture of oxygen and helium. These gases also dissolve under pressure in the diver's blood.

As the diver returns to the surface the air pressure decreases. This allows the dissolved gases to leave the blood as bubbles. If the diver resurfaces too quickly, the bubbles will be large and will block the blood vessels and cause damage to joints and delicate tissues like the brain. This is called 'decompression sickness'. To avoid this happening, a diver must resurface very slowly so the gases leave his blood as small harmless bubbles.

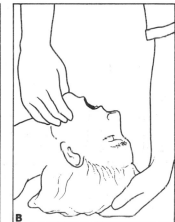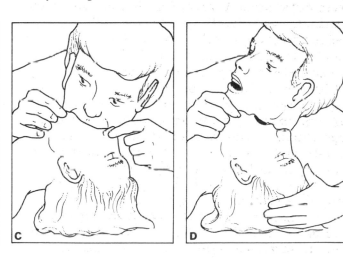

Fig. 2. Artificial respiration by mouth-to-mouth resuscitation. Lift the victim's head well back (**A** and **B**). This keeps the wind-pipe open. Pinch the victim's nose so that no air can escape through it, and inflate his lungs by gently blowing into his mouth (**C**). Repeat 12 times a minute.

40

Gaseous exchange

The body absorbs oxygen from the air and at the same time it releases into the air carbon dioxide produced by respiration. In other words the body 'exchanges gases' with the atmosphere. The technical name for this process is **gaseous exchange**, and it takes place within the alveoli of the lungs.

Absorption of oxygen

Blood entering the capillaries of the lungs is **deoxygenated**; that is, it contains no oxygen. This is because haemoglobin in its red cells has released all its oxygen to the cells of the body. But air breathed from the atmosphere into the alveoli is rich in oxygen. This oxygen dissolves in a film of water lining the alveoli (Figs. 1 and 2B), then diffuses through the alveoli and capillary walls, a distance of only 0·005 mm, into the blood. Here, it combines with haemoglobin in the red cells making oxyhaemoglobin.

The capillaries that cover the alveoli walls are narrower in diameter than red blood cells. The red cells are squeezed out of shape as they are forced through the lung capillaries by blood pressure. This helps oxygen absorption in two ways. First, it causes a large area of each red cell to be pressed against capillary walls through which oxygen is diffusing. Second, red cells are slowed down as they squeeze through the capillaries, thus increasing the time available for oxygen absorption. By the time blood leaves the lung capillaries it is oxygenated.

Release of carbon dioxide

Blood entering the lung capillaries is full of carbon dioxide which it has collected from respiring cells on its journey around the body. Most of the carbon dioxide is dissolved in the plasma either in the form of **sodium bicarbonate** or **carbonic acid**. A little carbon dioxide is transported by red cells as a substance called **carbamino-haemoglobin**.

As blood enters the lungs an enzyme in the red cells causes chemical reactions which break down sodium bicarbonate and carbamino-haemoglobin, releasing carbon dioxide gas. This gas diffuses through the capillary and alveoli walls into the film of water and then into the alveoli. Finally, it is removed from the lungs during breathing out.

Differences between breathed and unbreathed air

	Unbreathed air	Breathed air from a sleeping man	Breathed air from a running man
Nitrogen	78%	78%	78%
Inert gases	1%	1%	1%
Oxygen	21%	17%	12%
Carbon dioxide	Trace	4%	9%
Water vapour	Variable	Saturated	Saturated

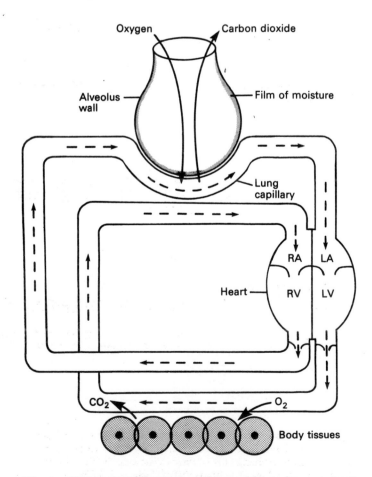

Fig. 1. Diagram of gaseous exchange (shown in more detail opposite).

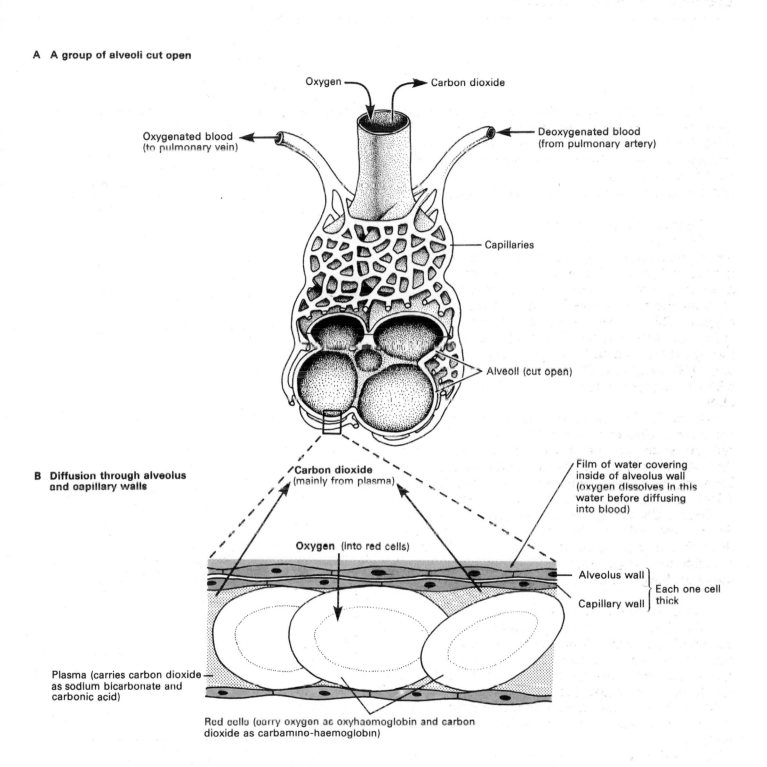

A   A group of alveoli cut open

Oxygen → ← Carbon dioxide

Oxygenated blood
(to pulmonary vein) ←

← Deoxygenated blood
(from pulmonary artery)

Capillaries

Alveoli (cut open)

**B   Diffusion through alveolus
and capillary walls**

Carbon dioxide
(mainly from plasma)

Film of water covering
inside of alveolus wall
(oxygen dissolves in this
water before diffusing
into blood)

Oxygen (into red cells)

Alveolus wall
Capillary wall

Each one cell
thick

Plasma (carries carbon dioxide
as sodium bicarbonate and
carbonic acid)

Red cells (carry oxygen as oxyhaemoglobin and carbon
dioxide as carbamino-haemoglobin)

Fig. 2.   Gaseous exchange is the absorption of oxygen from the air 'in exchange' for carbon dioxide which is released from the body. This exchange takes place in the alveoli of the lungs. Oxygen breathed into the lungs diffuses through a film of moisture lining the alveoli, and then passes through the alveoli and capillary walls into the red cells. Here it forms oxyhaemoglobin. Carbon dioxide is carried by blood mainly as sodium bicarbonate. An enzyme in red cells breaks this down releasing carbon dioxide gas. This diffuses into the alveoli and is breathed out of the lungs.

41

Homeostasis, excretion, and the urinary system

Homeostasis

Unit 34 explains that the cells of the body are bathed in a liquid called tissue fluid. This fluid comes from the blood; it supplies cells with food and oxygen and removes their waste products. Tissue fluid forms the environment in which cells live or, put another way, it forms the **internal environment** of the body.

A number of organs are constantly adjusting the temperature and contents of tissue fluid so that it is always as near perfect an environment as possible for the health, growth, and efficient functioning of cells. The organs which perform this task are said to be involved in **homeostasis**, or 'the maintenance of a constant internal environment'.

Organs concerned with homeostasis

The main organs of homeostasis are the lungs, skin, liver, and kidneys.

Lungs control the amount of carbon dioxide and oxygen in tissue fluid. They do this by removing carbon dioxide as fast as it is produced by respiration in cells, and by supplying cells with oxygen as fast as they use it.

The skin helps maintain the internal environment of the body at about 37°C. It does this by means of sweat glands, and other mechanisms described in Unit 43.

The liver and pancreas work together to maintain within precise limits the amount of glucose sugar in blood and tissue fluid. The liver also maintains amino acids and proteins at a constant level by breaking down any excess in a process called deamination, described in Unit 28. One product of deamination is a waste substance called urea, which is removed from the body by the kidneys.

Removal from the body of carbon dioxide by the lungs and urea by the kidneys are examples of excretion.

Excretion

Excretion is an extremely important part of homeostasis because it removes waste substances produced by the chemical reactions of metabolism, and substances which are excess to the body's requirements. If these substances were not removed, they would poison cells or slow down metabolism. The main organs of excretion are the lungs and the kidneys.

If carbon dioxide were not excreted by the lungs, it would accumulate in the body and form carbonic acid in quantities which would damage cells.

Water is an example of a substance which is often present in the body in excess of requirements. If, for instance, someone drinks more liquid than the body requires, the excess would dilute the blood and tissue fluid to a dangerous level if it were not excreted by the kidneys.

The most important type of excretion is the removal by the kidneys of substances produced by the deamination of excess amino acids. Deamination involves the production of ammonia. Ammonia is very poisonous, but the liver immediately converts it into harmless urea, and releases it into the blood. Kidneys extract urea from the blood and excrete it as part of a liquid called **urine**. Ammonia and urea contain nitrogen, therefore their removal from the body by the kidneys is called **nitrogenous excretion**. The kidneys are part of a set of organs called the **urinary system**.

The urinary system

The parts of the urinary system and their position in the body are illustrated in Figure 2.

Human kidneys are about 12 cm long by 7 cm wide, and are bean-shaped. A thin tube, the **ureter**, comes out of the concave side of each kidney and extends downwards to a single large bag called the **bladder**. The bladder has only one exit, a tube called the **urethra**, which leads to the body surface. The bladder end of the urethra is normally closed by means of rings of muscle (sphincters), which control the release of urine from the bladder.

Urine drains continuously out of the kidneys into the ureters, where it is forced downwards into the bladder by wave-like contractions of the ureter walls. The bladder stretches and expands in volume as it fills with urine, and when it is nearly full the stretching stimulates sensory nerve endings in its walls so that nerve impulses are sent to the brain. This is how you know when your bladder must be emptied. The sphincter muscles around the urethra are then voluntarily (consciously) relaxed to let urine drain from the bladder, through the urethra, and out of the body. This is called urination or, in technical terms, **micturition**.

A Skin

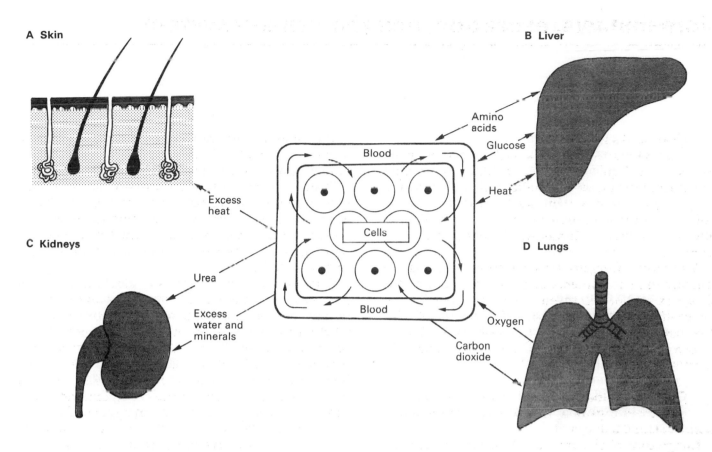

B Liver

C Kidneys

D Lungs

Blood

Amino acids

Glucose

Heat

Excess heat

Cells

Urea

Excess water and minerals

Blood

Oxygen

Carbon dioxide

Fig. 1. Diagram of homeostasis. A number of organs work non-stop at maintaining the temperature and contents of tissue fluid at constant levels. This activity is called homeostasis or: the maintenance of a constant internal environment. Lungs control the levels of carbon dioxide and oxygen; the liver controls levels of amino acids and sugar; the skin helps keep temperature at about 37 °C; and the kidneys remove waste substances such as urea, excess water, and mineral salts.

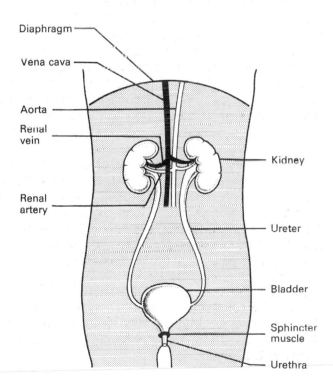

Diaphragm

Vena cava

Aorta

Renal vein

Renal artery

Kidney

Ureter

Bladder

Sphincter muscle

Urethra

Fig. 2. The urinary system. Kidneys produce a liquid called urine. It drains out of the kidneys and through the ureters which carry it to a muscular bag called the bladder. The bladder expands as it fills, and at a certain point nerve endings in its walls send impulses to the brain which signal that the bladder must be emptied. Sphincter muscles at the base of the bladder are relaxed and urine drains out of the body through the urethra.

42

Excretion by the kidneys

Each kidney receives blood at very high pressure through a **renal artery** (Fig. 2A). Inside the kidney this artery divides into capillaries with little loss of blood pressure. The capillaries carry blood to tiny cup-shaped structures called **Bowman's capsules**. Each Bowman's capsule is the beginning of a length of narrow tube 3 cm long called a **nephron** (Fig. 2C). There are over a million nephrons in each kidney, and their function is to produce urine.

Formation of urine

The capillary inside each Bowman's capsule divides into a tiny ball of inter-twined blood vessels called a **glomerulus** (Fig. 2C). Blood pressure in a glomerulus is so high that liquid is forced out through the capillary walls into the space inside the Bowman's capsule. This liquid is called **glomerular filtrate**, because it is formed by liquid from the blood filtering through capillary walls and the inner walls of the Bowman's capsules.

About 7·5 litres of glomerular filtrate are produced every hour. It contains urea to be excreted but, in addition, there are many useful substances in it such as glucose, amino acids, mineral salts, vitamins, and large amounts of water, which the body cannot afford to lose. The body does not lose these useful substances because they are reabsorbed into the bloodstream.

Reabsorption

Glomerular filtrate flows out of the Bowman's capsules into the tubular part of each nephron. It is here that reabsorption occurs. The walls of a nephron extract useful substances (glucose, etc.) from glomerular filtrate and pass them into blood flowing through capillaries surrounding the nephron (Fig. 2C). Removal of useful substances changes glomerular filtrate into **urine**. Normally, urine is made up of urea and small amounts of mineral salts dissolved in water. Urine drains out of the nephrons into collecting ducts, and then into ureters and out of the body as described in Unit 41.

Osmoregulation

Reabsorption does not merely save useful substances in danger of being lost from the body. It is also a means of regulating the amount of water and dissolved materials in blood and tissue fluid. The technical name for this process is **osmoregulation**, because it re-gulates the flow of water by osmosis between cells and the blood.

After a large meal, for example, or in a person suffering from diabetes, the amount of sugar in the blood begins to rise. At such a time the kidneys do not reabsorb all the glucose from glomerular filtrate, and so the excess passes out of the body in urine.

If large amounts of liquid are drunk the water content of the blood rises. Less water is then re-absorbed by the kidneys, and large amounts of dilute urine are produced. But if the body contains too little water the kidneys reabsorb a maximum amount from glomerular filtrate, leaving a small quantity of very concentrated urine. In hot weather the same process ensures that plenty of water is available for cooling the body by perspiration from the sweat glands.

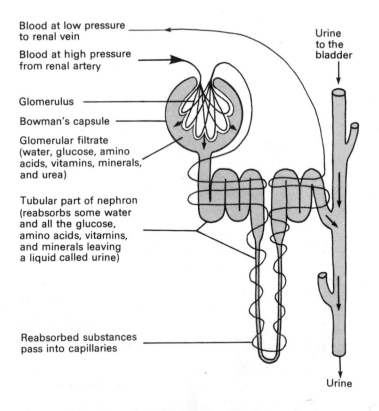

Blood at low pressure to renal vein

Blood at high pressure from renal artery

Glomerulus

Bowman's capsule

Glomerular filtrate (water, glucose, amino acids, vitamins, minerals, and urea)

Tubular part of nephron (reabsorbs some water and all the glucose, amino acids, vitamins, and minerals leaving a liquid called urine)

Reabsorbed substances pass into capillaries

Urine to the bladder

Urine

Fig. 1. Diagram of a nephron, showing how urine is formed. Blood pressure in a glomerulus forces a liquid called glomerular filtrate through the capillary walls into the Bowman's capsule. This filtrate contains urea, plus many useful substances which are reabsorbed into the blood by the tubular part of each nephron. This leaves urine, which drains out of the kidney.

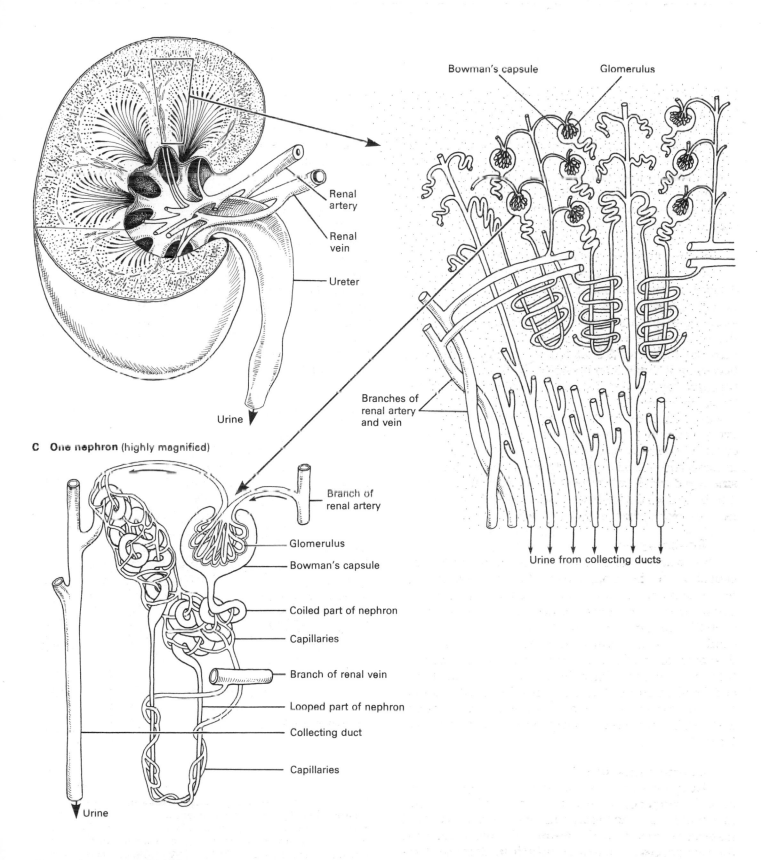

A **A kidney** (cut open)

Renal artery

Renal vein

Ureter

Urine

B **Part of a kidney** (magnified)

Bowman's capsule

Glomerulus

Branches of renal artery and vein

Urine from collecting ducts

C **One nephron** (highly magnified)

Branch of renal artery

Glomerulus

Bowman's capsule

Coiled part of nephron

Capillaries

Branch of renal vein

Looped part of nephron

Collecting duct

Capillaries

Urine

Fig. 2. Structure of a kidney.

43

Skin and temperature control

Humans, other mammals, and birds can maintain their body temperature at a fairly constant level despite temperature changes in their surroundings. In technical terms they are **homoiothermic**. All other animals are **poikilothermic**, which means their body temperature varies according to the temperature of their surroundings.

The temperature of mammals and birds is controlled automatically by involuntary (unconscious) mechanisms and also by voluntary (conscious) actions. These control temperature by maintaining a balance between heat lost from the body and heat produced inside it (Fig. 1).

Involuntary temperature control

The body automatically increases the rate at which it loses heat when conditions outside the body are near or above normal body temperature (36–37·5 °C), and when there is an increase in heat production within the body as occurs during vigorous exercise and illness. Heat is lost in the following ways.

Sweating As body temperature rises the sweat glands absorb liquid from tissue fluid and pour it onto the skin surface, where it evaporates and cools the body. It takes 2·43 kJ of heat energy to change 1 g of sweat into vapour. Therefore this amount of heat is lost from the body with every gram of sweat which evaporates from its surface.

The rate at which sweat evaporates from and therefore cools the body depends upon the amount of water vapour in the air around the body (humidity) and air movements (e.g. wind or fans). Sweat evaporates very quickly in hot dry conditions. But in hot humid conditions, especially in still air, sweat evaporates and cools the body very slowly or not at all. In these conditions body temperature may rise out of control, and if it reaches 41 °C or above the victim collapses and may die.

Rapid sweating resulting from strenuous exercise in hot climates may cause loss of up to 30 litres of water per day and 30 g of salt. Loss of water at this rate soon causes the blood to become thick and concentrated, so that it no longer circulates properly. Loss of salt causes muscle pains (heat cramp). In hot climates a person must not only drink a lot of fluids, but must also increase the amount of salt in the diet.

Strenuous physical work for long periods in a hot climate can result in a sudden failure of the ability to sweat. This is called **heat stroke**. If victims are not taken to a cool place immediately their temperature will rise uncontrollably with possibly fatal results.

Vasodilation This is the expansion of blood vessels. They become wider in diameter and so carry more blood. Whenever the body temperature begins to rise, vasodilation occurs in the arterioles which supply the dense network of capillaries just beneath the epidermis of the skin (Figs. 1 and 2A). As a result a large volume of over-heated blood flows very close to the body surface where it rapidly loses heat by radiation through the skin. This cools the body.

If the body begins to lose heat faster than it is producing it, the following events occur automatically.

Increased heat production In cold weather there is a general increase in the rate of metabolism. This results in the production of extra heat, especially from the liver. The appetite is also stimulated, which provides the extra fuel (food) needed to maintain a higher metabolic rate.

Vigorous exercise warms the body because it increases the rate of respiration and, thereby, heat production in muscle tissue. If the body starts to cool down in cold weather the muscles automatically begin jerky rhythmic movements. This is shivering. It warms the body by increasing the amount of heat which the muscles produce.

Reduction of heat loss In cold weather **vasoconstriction** occurs in the arterioles which supply skin capillaries. This means that the arterioles become smaller in diameter, which reduces blood flow near the body surface, thereby reducing to a minimum heat loss by radiation through the skin (Fig. 2B). Sweat glands cease to operate in cold weather. This reduces heat lost by evaporation but a small amount of water still evaporates through the skin from moist underlying tissues.

Voluntary temperature control

The skin contains sensory nerve endings sensitive to changes in its temperature. These nerve endings send impulses to the part of the brain where sensations of hot and cold are felt. People feel 'comfortable' when their skin is about 33 °C. If their skin falls below this temperature they feel 'cold' and may then switch on a heater, put on warmer clothing, indulge in vigorous exercise, etc., until they feel 'warm' again. When their skin rises above 33 °C, people will feel 'too hot' and may then open a window, put on lighter clothing, turn on a fan, etc.

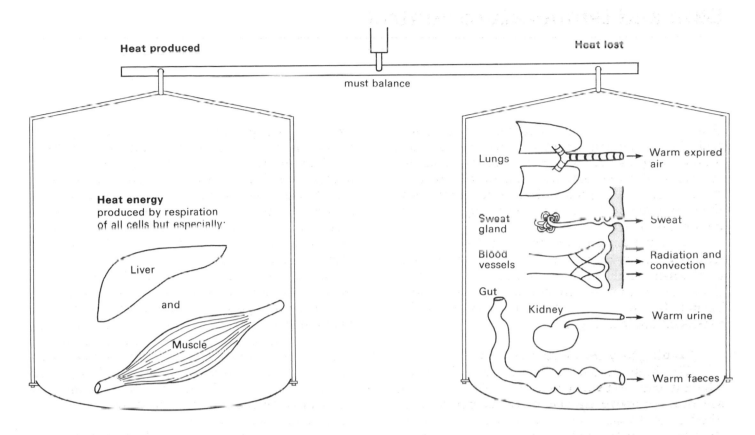

Fig. 1. There is a balance between heat loss and heat production. A constant body temperature at about 37 °C is maintained by mechanisms which achieve a balance between heat lost from the body and heat produced in the body.

A How heat is lost through the skin

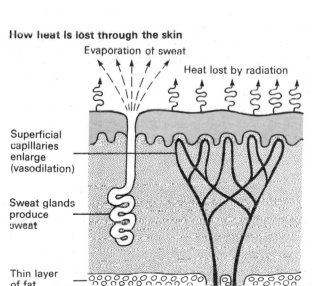

B How skin retains heat

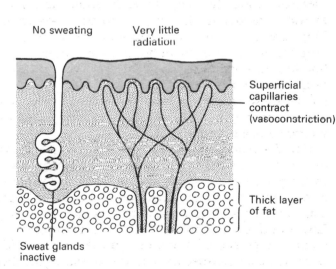

Fig. 2. Some temperature control mechanisms. When the body is too hot, sweat glands pour liquid onto the skin, where it evaporates and cools the body. In addition arterioles increase in diameter (undergo vasodilation) so that more blood flows through them into capillaries under the epidermis, where heat is lost by radiation. In cold weather sweat glands cease to work and arterioles contract (undergo vasoconstriction).

44

Structure and care of the skin and hair

Skin protects underlying tissues and organs from dirt, germs, injury, and the harmful effects of strong sunlight. It is waterproof, controls the rate at which water is lost by evaporation from the body, and helps regulate body temperature in ways described in Unit 43. The skin contains several types of sensory nerve endings, described in Unit 45.

There are two main layers to the skin: an upper layer called the **epidermis** and a lower one called the **dermis** (Fig. 1).

Epidermis

The epidermis itself is made up of several layers. The lowest is called the **germinative layer**. It produces all the other layers by cell division. As new cells are produced they push older cells above them towards the surface of the skin. There are no blood vessels in the epidermis, and so the cells die from lack of oxygen and food as they are pushed further away from the germinative layer. As the cells die they fill with tiny granules of the protein **keratin** and form the **granular layer** of the epidermis. Keratin eventually makes the cells hard and dry. They change into thin dry flakes which form the tough **cornified layer** of the skin. This covers the outer surface of the body. It is constantly removed by washing and as the skin rubs against objects, but is replaced by growth from below. If the skin is cut, grazed, or burned it is repaired with new cells produced by the undamaged germinative layer surrounding the wound.

Dermis

The dermis contains many connective tissue fibres. Wavy bands of tough collagen fibres limit the extent to which skin can be stretched, and elastic fibres pull the skin back into shape after it has been stretched.

There are blood vessels in the dermis, particularly in the many dome-shaped **dermal papillae** below the epidermis (Fig. 1). Blood in these papillae nourishes the germinative layer and plays an important part in temperature control. In many places the germinative layer grows down into the dermis forming sweat glands and hair follicles.

Sweat glands Adults have over two million sweat glands. Each consists of a coiled tube which opens at a pore on the skin surface. Cells lining the tube extract liquid from surrounding capillaries.

This liquid moves up the tube onto the skin where it evaporates. There are two types of sweat gland. One type, found everywhere in the skin except the lips, produces a watery solution of salt and a little urea which evaporates from the skin cooling the body. A second type, found under the arms and between the legs, produces a thicker liquid which can give rise to 'body odour' described below.

Hair follicles These are deep pits in which hairs grow. The shaft of a hair is formed at the base of a follicle by a root consisting of capillaries and germinative cells. Cells of a hair shaft quickly fill with keratin and die. Every three or four years hairs on the head stop growing and fall out. Soon afterwards however the hair is replaced by new growth from the same root. A tiny erector muscle is attached to the base of each follicle. In humans these muscles have no function, but in hairy mammals they contract in cold weather, making all the hairs stand upright.

At one side of each follicle there is a flask-shaped **sebaceous gland**. These glands produce an oily liquid called **sebum** which spreads over the hairs and skin, making them supple and waterproof and helping to prevent growth of bacteria.

Care of skin and hair

If the skin and hair are not washed regularly they collect dirt, dried sweat (particularly under the arms), and loose cells from the cornified layer of the skin. These form an ideal breeding ground for lice, mites, and fleas, and for bacteria and fungi. The bacteria decompose the dirt and dried sweat producing an unpleasant smell called body odour. But, more important, fungi which grow on dirty skin can cause serious illness. Athlete's foot, for example, is caused by a fungus which attacks soft skin between the toes and in bad cases spreads over the whole foot. The skin peels off leaving very sore patches. Furthermore, if dirty skin is cut the resulting wound is quite likely to become infected with germs.

Particular attention should be paid to washing the hands and nails as these can carry germs which may be transferred to food and passed to other people. Combs and hair brushes should also be washed regularly as they can re-infect hair which has just been washed clean.

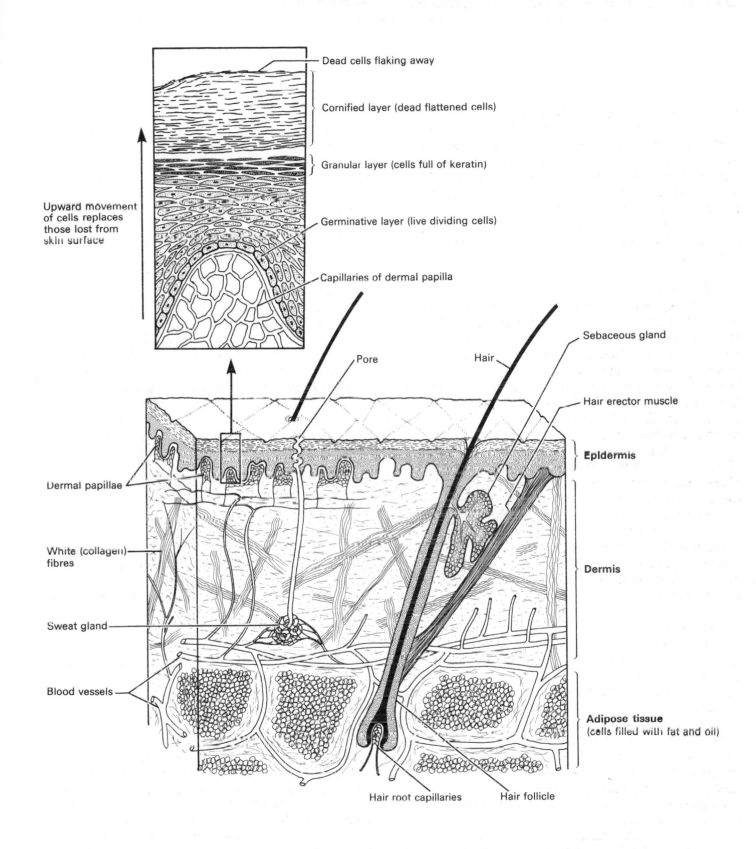

Dead cells flaking away

Cornified layer (dead flattened cells)

Granular layer (cells full of keratin)

Germinative layer (live dividing cells)

Upward movement of cells replaces those lost from skin surface

Capillaries of dermal papilla

Pore

Hair

Sebaceous gland

Hair erector muscle

Epidermis

Dermal papillae

White (collagen) fibres

Dermis

Sweat gland

Blood vessels

Adipose tissue (cells filled with fat and oil)

Hair root capillaries

Hair follicle

Fig. 1. Structure of the skin (sensory nerve endings have been omitted). Skin consists of two layers: an epidermis and a dermis. Skin protects the body from dirt, germs, strong sunlight, and injury. It is waterproof, helps regulate body temperature, and contains many sensory nerve endings. It repairs itself when damaged and replaces itself as fast as it is worn away.

89

45

Senses in the skin, tongue, and nose

Skin

There are at least five different types of sensory nerve ending in the skin. These, and all other sensory nerve endings, are called **receptors**, because they are the parts of the nervous system at which information is received. It is not yet certain how skin receptors work or what their individual functions are. Nevertheless they have been named according to their most likely functions, which are sensitivity to: touch, pressure, pain, temperature, and hair movements (Fig. 1).

Touch   These receptors are located immediately below the epidermis, and are most numerous in the surface of the tongue and fingertips. Touch receptors are stimulated by light pressure on the skin and enable a person to distinguish between textures such as rough, smooth, hard, and soft. A different type of touch receptor detects hair movements caused by wind or objects brushing against the skin.

Pressure   These receptors are situated beneath the dermis and are stimulated by heavy pressure.

Pain   These receptors consist of branched nerve endings in the epidermis and dermis. They are quite evenly distributed throughout the skin and occur almost everywhere inside the body except the brain. Pain is unpleasant but very important because it is a warning signal that something is going wrong inside the body, or that some part of the skin is being damaged.

Temperature   There appear to be separate 'heat' and 'cold' receptors in the skin. They are stimulated by sudden changes in temperature.

Nose

Smell receptors, or olfactory organs, are sensitive to chemicals in the air (Fig. 2). These chemicals must first dissolve in a film of moisture which covers the receptors.

Tongue

Taste receptors, or taste buds, in the tongue are also sensitive to chemicals (Fig. 3B). All taste buds look alike (Fig. 3C) but in fact there are four different types: those which respond to salt, sweet, sour, and bitter-tasting substances. Groups of each type of receptor are concentrated in certain areas of the tongue (Fig. 3A). The different flavours of food and drink are identified according to how much they stimulate these four types of receptor.

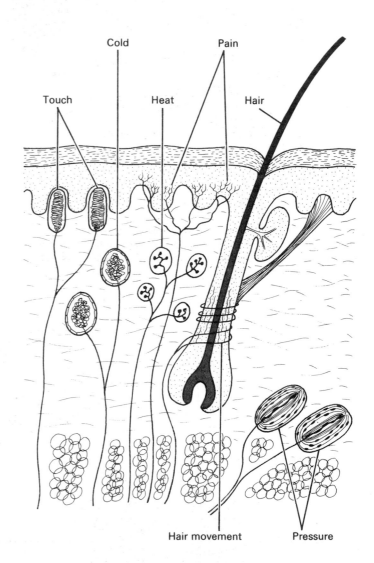

Fig. 1.   Sensory nerve endings (receptors) in the skin.
Touch receptors are sensitive to light pressure. They enable us to distinguish between different textures. Pressure receptors are under the dermis and respond to heavy pressure. Pain receptors are branched nerve endings in the epidermis and most other parts of the body. Heat and cold receptors respond to changes in temperature. Hair movements are detected by nerves wrapped around the hair roots.

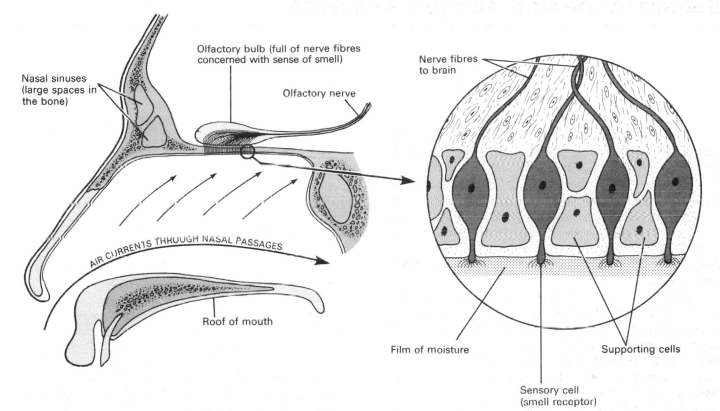

Fig. 2. Smell receptors are called olfactory organs. They are situated in the roof of the nasal passages and detect chemicals in the air. First the chemicals must dissolve in a film of moisture which covers the receptors.

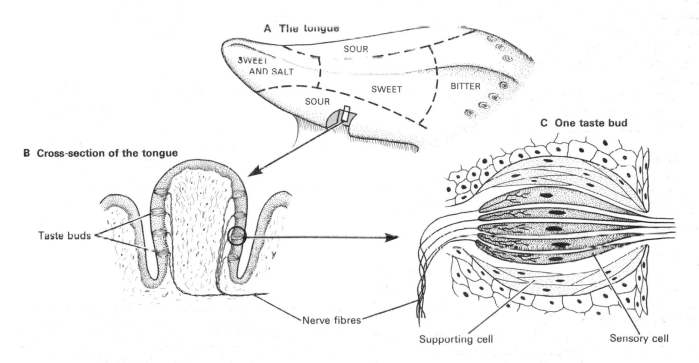

Fig. 3. There are four types of taste bud, those which respond to: salt, sweet, sour, and bitter tastes. Groups of each type are concentrated in certain areas of the tongue.

46

Structure of the eye

Structures around the eyes

The eyes are situated in cavities in the skull called **orbits** (Fig. 12.1). These enclose and protect all but the front of the eyes.

Eyelids   The exposed front of each eye is protected by a thin skin called the **conjunctiva**, and by the eyelids, eyelashes, and tear glands (Fig. 1). When the eyes are open, the eyelashes form a protective net which traps large particles of dust. The tear glands are situated above the eyeballs. They produce a continuous stream of watery liquid which is wiped over the conjunctiva every few seconds by automatic (reflex) blink movements of the eyelids. Tears keep the conjunctiva moist, wash it clean, and contain an enzyme which destroys bacteria. When dust or chemicals blow into the eyes, the flow of tears and blinking rate increase automatically until the eyes are cleaned.

Eye muscles   Each eye is held in place and moved within its orbit by six **extrinsic muscles** attached to the outer surface of the eyeball and the skull. These muscles can rotate the eyes to follow moving objects, and direct the gaze to a chosen object. The movements are precisely co-ordinated so that both eyes work together and are always directed at the same spot.

Parts of the eyeball

Sclerotic layer   This is a thick layer of tough white fibres which form the outer wall of the eyeball. The sclerotic protects the delicate inner structures and helps maintain the eye's rounded shape.

Cornea   This is a transparent circular window at the front of the eyeball. It is continuous with the sclerotic. The curved shape of the cornea causes light rays to bend (be reflected) as they pass through it, so that they converge on the back of the eyeball.

Choroid layer   This is a layer of blood vessels and cells full of dark brown pigment. It lines the inside of the sclerotic at the back of the eye. The blood vessels supply the eye with food and oxygen, and the pigment prevents light from being reflected around inside the eyeball. The front edge of the choroid forms a thick ring of tissue called the ciliary body.

Ciliary body   This consists of blood vessels, and a mass of muscle fibres which make up the **ciliary muscle**. This muscle controls the shape of the lens during focusing, as described in Unit 47.

Fibres called the **suspensory ligaments** run from the inner edge of the ciliary body to the outer rim of the lens. These fibres hold the lens in place and play a part in focusing.

The lens   This consists of layers of transparent tissue, arranged like the skins of an onion, which are enclosed in a transparent membrane. The lens is concerned with focusing a clear image on the back of the eye. The cornea alone can do this, but the lens makes it possible for the image to be refocused during shifts of vision between near and distant objects.

Iris   This is the coloured part of the eye. It is situated in front of the lens and has a hole at its centre called the **pupil**. The iris is made up of radial and circular muscles. When the radial muscles contract the pupil is enlarged, and when the circular muscles contract the pupil is reduced in size. In this way the iris regulates the amount of light entering the eye: it opens the pupil in dim light and reduces it to pin-hole size in bright light.

Retina   This is a layer of sensory nerve endings which cover the back of the eye. The cornea and lens focus light rays on the retina forming a clear, upside-down, full-colour image of the outside world. The nerve endings in the retina are sensitive to light and respond to this image by sending nerve impulses along nerve fibres which extend from the eye and along the optic nerve to the brain. The retina and image formation are described in Unit 47.

Aqueous and vitreous humours   The space in front of the lens is filled with a liquid called aqueous humour, and the space behind the lens is filled with a jelly called vitreous humour. Aqueous humour is similar to blood plasma. It is produced by blood vessels in the ciliary body, and carries food and oxygen from this blood supply to the cornea and lens. It also inflates the front of the eye, thereby maintaining the curved shape of the cornea. Aqueous humour drains out of the eye into the bloodstream through a tiny canal in the angle between the iris and cornea.

Vitreous humour gives shape and firmness to the rear compartment of the eye, and keeps the retina in contact with the choroid and sclerotic layers. Both the aqueous and vitreous humours play a part in focusing an image on the retina by bending light rays passing through them.

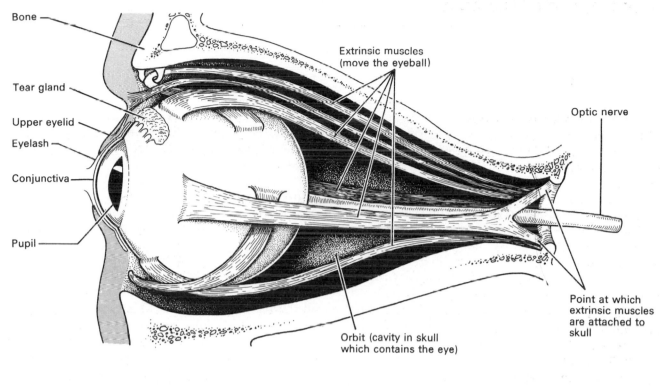

Bone

Tear gland

Upper eyelid

Eyelash

Conjunctiva

Pupil

Extrinsic muscles
(move the eyeball)

Optic nerve

Point at which
extrinsic muscles
are attached to
skull

Orbit (cavity in skull
which contains the eye)

Fig. 1. Structures attached to the eyeball. Eyes are protected by sockets in the skull called orbits. The front of each eye is protected by two eyelids, eyelashes, and a tear gland.

Tears keep the eyes moist, wash away dirt, and contain an enzyme which destroys bacteria. Eyes are held in place and moved inside their sockets by extrinsic muscles.

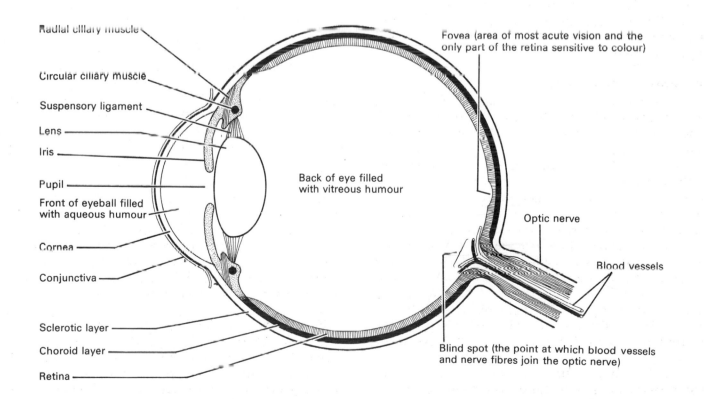

Radial ciliary muscle

Circular ciliary muscle

Suspensory ligament

Lens

Iris

Pupil

Front of eyeball filled
with aqueous humour

Cornea

Conjunctiva

Sclerotic layer

Choroid layer

Retina

Fovea (area of most acute vision and the
only part of the retina sensitive to colour)

Back of eye filled
with vitreous humour

Optic nerve

Blood vessels

Blind spot (the point at which blood vessels
and nerve fibres join the optic nerve)

Fig. 2. Horizontal section through the right eye.

93

47

Focusing and structure of the retina

Focusing (accommodation)

The technical name for refocusing the eye between near and distant objects is accommodation. This is accomplished by changing the shape of the lens, and is possible because the lens is made of elastic material. The shape of the lens is changed by contraction of the radial and circular ciliary muscles.

Accommodation for a distant object An eye is focused on a distant object by changing the lens to a flattened (less convex) shape. This reduces to a minimum the power of the lens to bend (refract) light. A flattened shape is needed to focus the almost parallel light rays from distant objects (Fig. 1A). The lens is flattened by contraction of the **radial ciliary muscles**. These pull against the **suspensory ligaments**, which pull against the lens stretching it into a flatter shape.

Accommodation for a near object An eye is focused on a near object by changing the lens to a rounded (more convex) shape. This increases the lens's power to refract light, which is necessary in order to focus the diverging light rays from a near object onto the retina (Fig. 1B). The lens is made more rounded by contraction of the **circular ciliary muscles** and relaxation of the radial ones. Circular ciliary muscles contract to form a circle with a smaller diameter. This reduces tension on the suspensory ligaments and allows the lens to become rounded in shape.

When a lens system is focused on a near object, its depth of focus is low. This means that a narrow zone, perhaps only a few millimetres wide, is clearly focused. However, when the eyes focus on a near object there is simultaneous contraction of the pupils, which increases the depth of focus.

The retina

The retina contains light-sensitive receptors of two types: about 125 million are called **rods**, and about 6 million are called **cones**, because of their shape.

Rods and cones do not face the light. Over most of the retina they are buried first under a layer of the nerve fibres which conduct impulses from the retina to the brain, and second under a layer of capillaries (Fig. 2B). But this is not true of an area of retina called the **fovea**. The fovea is a small depression in the retina immediately opposite the lens. There are no capillaries over this region and the layer of nerve fibres is very thin (Fig. 2A).

Rods and cones are not evenly distributed over the retina. The fovea consists exclusively of cones, packed tightly together with 147 million/mm². Elsewhere the retina consists mostly of rods.

Groups of up to 150 rods are connected to the brain by only one nerve fibre, whereas each cone has either its own exclusive nerve fibre connection to the brain or it shares one with only a few others.

This means that an image falling on the fovea, where there are only cones, produces a much clearer visual impression in the brain than an image falling on the rest of the retina. This is because the part of an image which falls on the fovea is minutely analysed by its tightly packed cones, each of which sends separate impulses to the brain. In comparison, the image which falls outside the fovea is analysed by large patches of rods which share a single fibre connection to the brain. When someone wishes to examine something carefully, they automatically move their eyes or the object until its image falls on the fovea.

Rods at the outermost edge of the retina do not produce a visual impression in the brain. They serve only to trigger reflexes which turn the eyes towards objects just beyond the limits of normal vision. This is what happens when something is seen 'out of the corner of the eye'.

There are several other important differences between rods and cones. Rods continue to work in dim light, but they are not sensitive to colours. Cones, on the other hand, work only in bright light and are sensitive to colours. These facts explain the differences between day and night vision. Daylight vision is in colour, and has precise detail because it results mostly from images which fall on the cones of the fovea. However, the cones stop working as daylight fades, and vision relies more and more on the rods. Towards evening, a person loses the ability to see colours, and vision becomes less distinct.

The **blind spot** is a point in the retina where blood vessels and nerve fibres leave the eye and form the **optic nerve** (Fig. 46.2). The blind spot contains neither rods nor cones therefore, as its name implies, it is entirely insensitive to light. The blind spot is described and demonstrated in Unit 48.

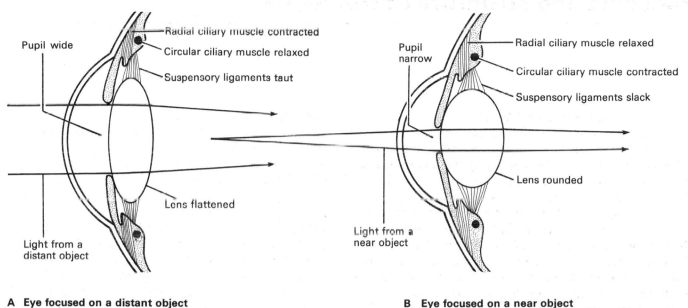

A Eye focused on a distant object
The radial ciliary muscles contract and pull against the suspensory ligaments. This stretches the lens into a flattened (less convex) shape

B Eye focused on a near object
Circular ciliary muscles contract. This releases tension on the suspensory ligaments and the lens becomes more rounded in shape

Fig. 1.   Diagram of changes in the eye during focusing.

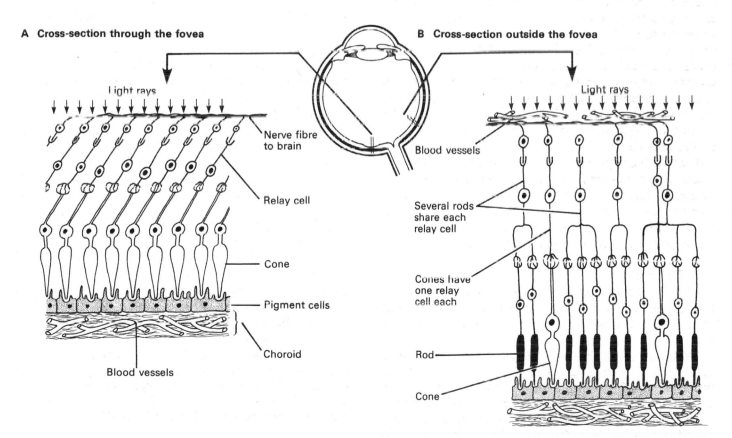

Fig. 2.   Structure of the retina. It contains light-sensitive receptors called rods and cones. Cones are concentrated in the fovea. They work only in bright light and are sensitive to colour. Rods work in dim light but are not sensitive to colour.

48

Binocular vision and eye defects

Binocular vision

There are four main advantages to having two eyes, or binocular vision. First, two eyes give a larger field of vision than one (check this by comparing what you can see with one eye and with two). Second, a defect in one eye does not result in blindness provided the area which cannot be seen with one eye is seen clearly by the other. This explains why we are unaware of the blind spot in each eye (Figs. 1 and 46.1). A third advantage of binocular vision is that it allows distance judgements to be made, as explained below. Fourth, animals with two forward-facing eyes (humans, apes, owls, etc.) possess stereoscopic or three-dimensional vision.

Stereoscopic vision   This type of vision depends upon having two eyes which can look at the same object. In humans the eyes are about 6·3 cm apart, and when both are focused on an object they each see a slightly different view (Fig. 2). This can be demonstrated by closing one eye at a time and comparing the view from each. The visual centre in the brain combines these two views and makes from them a three-dimensional impression. This gives a good idea of the size and shape of objects.

Distance judgements   The brain can judge distances up to 50 metres with accuracy by comparing the image from each eye. Beyond this distance objects produce an almost identical image in both eyes. Distance judgements are also aided by stretch receptors in the eyes. These detect changes of tension in the ciliary muscles during accommodation and in the extrinsic muscles as they swivel the eyes inwards or outwards to look at an object.

Eye defects

Old sight, or presbyopia   The lens continues to grow throughout life, but at a very slow rate after adolescence. By the age of about sixty, the centre of the lens is so far from supplies of oxygen and food that its cells die. This process reduces the lens's elasticity and its power to change in shape. It then becomes more or less fixed into a shape suitable only for distant vision. Therefore, old people usually require 'reading glasses' which have converging lenses to give their eyes extra power for close work.

Long sight, or hypermetropia   Long sight occurs when the distance between the lens and the retina is shorter than normal (Fig. 3B). An image cannot be focused in so short a distance; in fact the point of clear focus is somewhere behind the retina. Long sight can be corrected by fitting spectacles with converging lenses, which add to the refractive power of the eye.

Short sight, or myopia   One cause of short sight is an abnormally elongated eyeball (Fig. 3C). That is, one in which the distance between the lens and the retina is so great that the point of clear focus is in front of the retina. This defect can be corrected by fitting spectacles with diverging lenses, which reduce the refractive power of the eye.

Astigmatism   This is a condition which occurs when the cornea lacks a perfectly dome-shaped surface. It is then impossible for the eyes to focus simultaneously on both vertical and horizontal lines (Fig. 3B). Astigmatism can be corrected by fitting spectacles with cylindrical lenses.

D Astigmatism
Occurs when the shape of the cornea is deformed. It is not possible to see horizontal and vertical lines clearly at the same time. Astigmatism is corrected with a cylindrical lens.

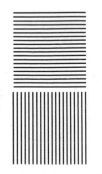

Fig. 1.   Blind spot experiment. Hold this figure at arm's length. Close your left eye and stare at the cross with your right eye. Note that the black circle is still visible. Bring the book slowly towards your face. At a certain point the circle will disappear. This happens when its image falls on the blind spot of your right eye. Why are you normally unaware of your blind spots?

Fig. 3 continued

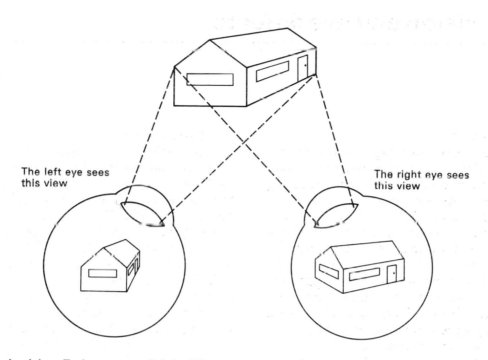

The left eye sees
this view

The right eye sees
this view

Fig. 2. **Stereoscopic vision.** Each eye sees a slightly different view. The brain puts these two views together to make a three-dimensional impression.

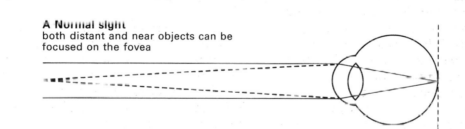

A Normal sight
both distant and near objects can be focused on the fovea

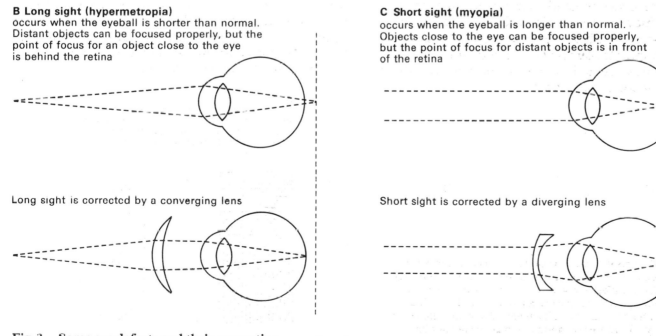

B Long sight (hypermetropia)
occurs when the eyeball is shorter than normal. Distant objects can be focused properly, but the point of focus for an object close to the eye is behind the retina

Long sight is corrected by a converging lens

C Short sight (myopia)
occurs when the eyeball is longer than normal. Objects close to the eye can be focused properly, but the point of focus for distant objects is in front of the retina

Short sight is corrected by a diverging lens

Fig. 3. **Some eye defects and their correction.**

49

The ear and hearing

Sound waves travelling through the air are collected by the funnel-shaped **pinna** of each ear (Fig. 2). The pinna directs sound waves down a short tube, the end of which is closed off by a sheet of skin and muscle called the **ear drum**. Sound waves make the ear drum vibrate in and out.

The structures described so far are part of the **outer ear**. Behind the ear drum is an air-filled space called the **middle ear**. A passage called the **Eustachian tube** connects this space with the back of the mouth. The Eustachian tube opens during swallowing, letting air in or out of the middle ear so that air pressure is always the same on both sides of the ear drum.

A chain of three tiny bones called the **ear ossicles** connects the ear drum with another sheet of skin called the **oval window**, which covers a tiny hole in the bones of the skull opposite the ear drum. When the ear drum vibrates, the ear ossicles move against each other in such a way that they lever the oval window in and out. These movements cause vibrations to pass into the inner ear.

The **inner ear** is a complicated series of tubes in the bones of the skull. The part of the inner ear concerned with hearing is called the **cochlea**, and is coiled like the shell of a snail (Fig. 2 A and B).

The cochlea is actually three tubes in one. The two outer tubes contain a liquid called **perilymph** and the middle tube contains **endolymph** (Fig. 2C). The floor of the middle tube is called the **basilar membrane**. This membrane supports a layer of sensory nerve endings called **hair cells** which stand upright like the pile of a carpet. The tips of the hairs are embedded in another membrane which overhangs them like a shelf jutting out of a wall.

Movements of the oval window cause vibrations (pressure waves) in the perilymph of the cochlea. These vibrations spread into the endolymph and finally cause the 'carpet' of hair cells to move up and down. This movement stimulates them to send nerve impulses along the auditory nerve to the brain, where they produce the sensation of sound (Fig. 1).

There is a theory that high-pitched notes cause vibrations only in hair cells nearest the oval window, while low-pitched notes cause vibrations further up the cochlea spiral. If this is true, the brain must distinguish notes from each other according to the region of hair cells stimulated as well as by the frequency of their vibrations. Whatever the mechanism it must

be extremely sensitive, for humans not only distinguish between very similar notes at a wide range of different volumes, but between the same note played on many different instruments. The response of the ear to high-pitched sounds decreases with age and so old people gradually lose their ability to hear high notes.

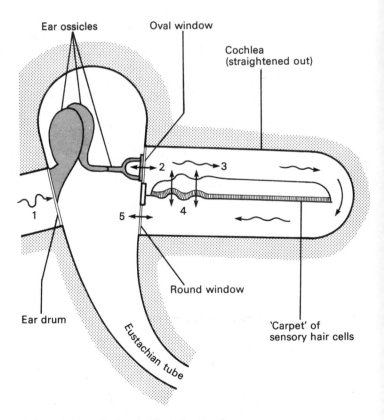

1. Sound waves in the air cause the ear drum to vibrate in and out

2. This causes the ossicles to vibrate and the stapes moves the oval window in and out

3. This sets up pressure waves in the cochlea

4. These pressure waves vibrate the sensory hair cells which send nerve impulses to the brain, causing the sensation of hearing

5. The pressure waves travel around the cochlea and move the round window in and out

Fig. 1.   Diagram of how the ear changes sound waves into nerve impulses. (The cochlea is shown as a straight tube for the sake of simplicity.)

A Structure of the ear (the middle and inner ears are diagrammatic and drawn on a larger scale than the outer ear)

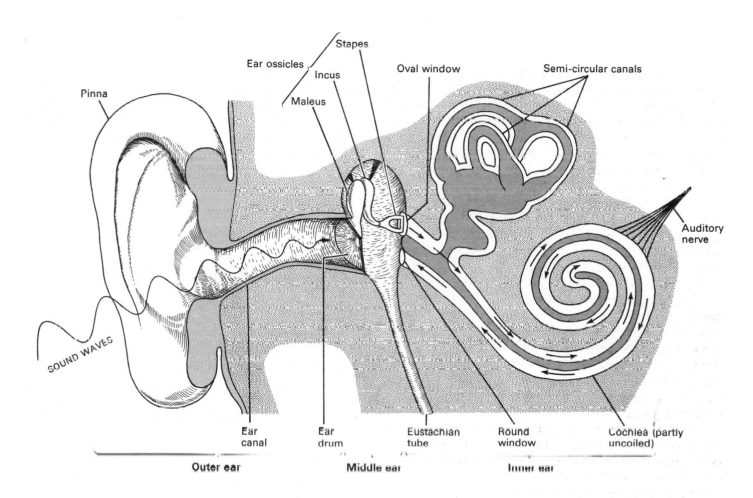

Pinna

Ear ossicles

Stapes

Incus

Maleus

Oval window

Semi-circular canals

Auditory nerve

SOUND WAVES

Ear canal

Ear drum

Eustachian tube

Round window

Cochlea (partly uncoiled)

Outer ear

Middle ear

Inner ear

B Side view of cochlea (showing spiral shape)

C Cross-section through the cochlea

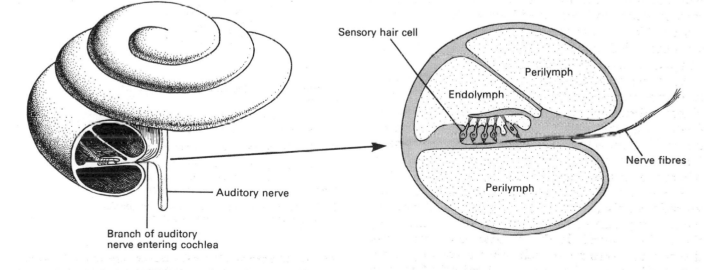

Auditory nerve

Branch of auditory nerve entering cochlea

Sensory hair cell

Perilymph

Endolymph

Nerve fibres

Perilymph

Fig. 2.   Parts of the ear.

50

Sense of balance

Many different sense organs play a part in helping us to maintain our balance. The eyes supply information about the position and movements of the body, and so do hundreds of stretch receptors in the muscles and joints. The parts of the inner ear other than the cochlea are also concerned with balance. The **semi-circular canals** detect changes in the direction of movements; and the **utricles** and **saccules** detect changes in the speed of movements (acceleration and deceleration) and changes in the position of the body (whether it is upright or tilted).

Semi-circular canals

These are curved tubes surrounded by perilymph and filled with endolymph (Fig. 1). At one end of each semi-circular canal is a swelling called an **ampulla**. These contain sensory hair cells. The hairs of these cells are embedded in a cone of jelly called a **cupula** (Fig. 2A). When the head moves it causes endolymph in one or more of the canals to press against a cupula and bend it over (Fig. 4). This stretches the sensory hair cells which send impulses to the brain. The brain calculates the direction of the movement from the frequency of nerve impulses it receives from each ampulla.

Utricles and saccules

These are spaces in the skull filled with endolymph (Figs. 1 and 2). There is a patch of sensory hair cells on the inner surface of each utricle and saccule (Fig. 2). The hairs of these cells are embedded in a mass of jelly containing tiny pieces of chalk called **otoliths**. Together, the hair cells, jelly, and otoliths are called a **sensory macula** (Figs. 2B and 3). The otoliths add considerably to the weight of the jelly. When the head is upright and still, the otoliths press down on the hair cells, but if the head tilts or is suddenly accelerated or decelerated, the otoliths pull against the hair cells causing them to send nerve impulses to the brain (Fig. 3). The brain calculates the angle of tilt or change in speed from the frequency of impulses which it receives from each utricle and saccule.

When the body is still few impulses reach the brain from the ampullas and maculas, but when the body is tilted and swayed as happens on a ship in rough seas, the brain receives a barrage of impulses from these organs. This over-stimulation may upset other areas of the brain and cause sea sickness.

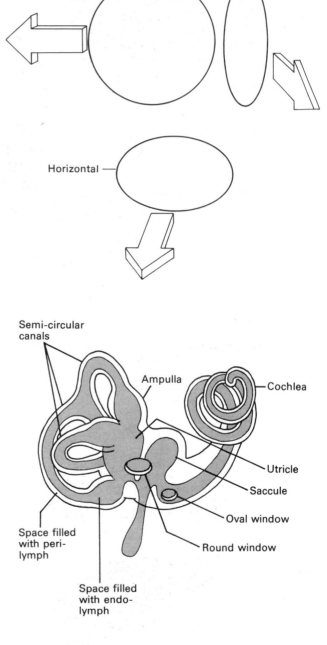

Fig. 1. The inner ear. This is also called the labyrinth because it consists of a complex series of passageways embedded in bone. The passages concerned with balance are the semi-circular canals, the utricles, and the saccules. Semi-circular canals detect changes in the direction of movements. Utricles and saccules detect acceleration, deceleration, and tilting of the head.

A Structure of an ampulla

B Structure of a macula

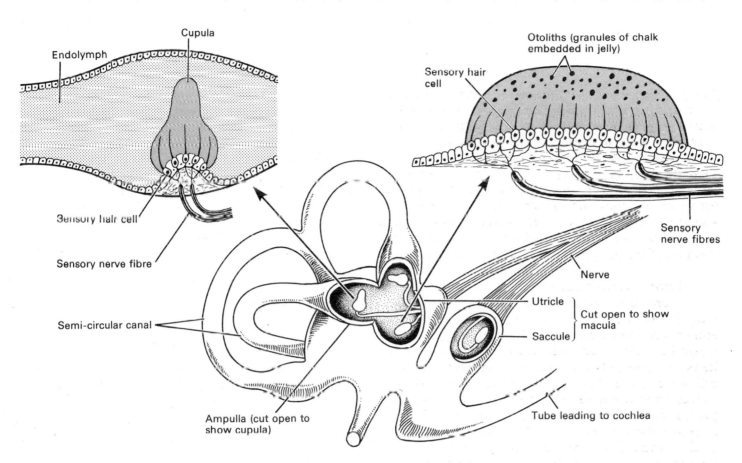

Fig. 2. **An ampulla and utriculus cut open to show the position of the cupula and macula.**

Head level

Head tilted

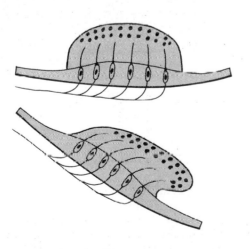

Fig. 3. **How a macula works.** When the head is level the otoliths press downwards. When the head tilts, speeds up, or slows down, the otoliths pull against the hair cells which send impulses to the brain.

Fig. 4. **How semi-circular canals work.** When the head moves one way endolymph is 'left behind' and apparently flows the other way, bending the cupula. This stimulates the hair cells which send impulses to the brain.

101

51

Parts of the nervous system

Nervous tissue and nerve impulses are described in Unit 9. This Unit and the next two describe the nervous system as a whole.

The nervous system consists of the brain, the spinal cord, and thousands of nerves which connect these two with all parts of the body. The brain and spinal cord form the **central nervous system**. Nerve fibres outside this system occur in bundles called nerves which, together, form the **peripheral nervous system**. **Cranial nerves** are those which are connected directly to the brain. Examples are nerves from the eyes and ears. **Spinal nerves** arise from the spinal cord (Fig. 1B).

The skull and backbone enclose and protect the central nervous system. It is also protected by two membranes called **mininges**, which occur between the bone and nervous tissue (Fig. 1A). The outer of the two meninges is a thick fibrous mat, while the other consists mainly of blood vessels. In between the meninges there is a space filled with **cerebro-spinal fluid**. This fluid acts as a cushion and also provides nervous tissue with food and oxygen. The same fluid also fills the hollow interior of the central nervous system.

The nervous system co-ordinates the activities of the body. Conscious muscular activities are co-ordinated using information from the eyes, ears, nose, and other external senses. Unconscious activities such as heart-beat and breathing rates are co-ordinated using information from internal sense organs which detect changes in blood pressure, body temperature, levels of carbon dioxide and glucose in the blood etc. Unconscious activities are controlled by the autonomic nervous system, which is described in Unit 53.

The brain

The human brain weighs about 1.5 kg and contains thousands of millions of neurones. Each neurone has synapses with a thousand or more other neurones, making an immensely complex network of cells and nerve fibres. The 'lowest' part of the brain, that is, the part which merges with the spinal cord, is called the medulla oblongata.

Medulla oblongata This region is concerned with many unconscious processes including the regulation of blood pressure, body temperature and rates of heart-beat and breathing. It performs these tasks through connections with the autonomic nervous system. The medulla also contains the mass of nerve fibres which connect the brain and spinal cord. Above the medulla is the cerebellum.

Cerebellum This receives impulses from the semi-circular canals and from stretch receptors in the muscles and joints. It uses information from these sources to maintain muscle tone and, thereby, a balanced posture, and it also co-ordinates muscles during activities like walking, running, dancing, and riding a bicycle. Above the cerebellum is the cerebrum.

Cerebrum This is the largest part of the human brain. It is a dome-shaped mass of nervous tissue made up of two halves called the **cerebral hemispheres**.

The cerebrum consists of an outer layer 1 mm thick of neurone cell bodies (grey matter) and a much thicker inner layer of nerve fibres (white matter). The outer layer is called the **cerebral cortex**, or **cortex** for short. This is the region where the main functions of the cerebrum are carried out.

The cortex is concerned with all forms of conscious activity: sensations such as vision, touch, hearing, taste, and smell; control of voluntary movements; reasoning, emotion; and memory.

It is possible to draw a map of the cortex showing the functions of certain areas (Fig. 2).

Functional areas of the cortex The cortex of each cerebral hemisphere has a number of **sensory areas**. These receive impulses from sense organs all over the body. There are separate sensory areas for vision, hearing, touch, taste, and smell, and it is in these areas that sensations or 'feelings' occur. Other regions of the cortex are called **motor areas**, because they have connections through motor nerve fibres with voluntary muscles all over the body. Separate motor areas control muscles of the limbs, abdomen, neck, tongue, etc.

Finally, there are relatively unchartered regions of cortex called **association areas**. It is here that 'association' takes place between information from the sensory areas and remembered information from past experiences. Conscious thought then takes place and decisions are made which often result in conscious muscular activity controlled by the motor areas (Fig. 2).

A Diagrammatic cross-section through the skull meninges and brain

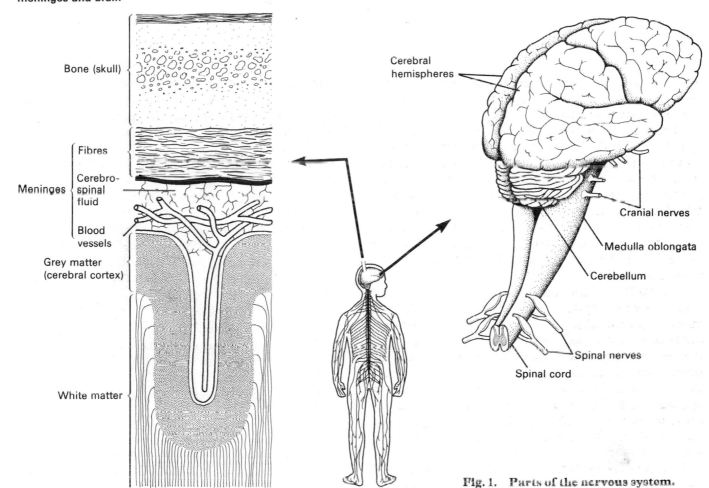

Bone (skull)

Meninges
- Fibres
- Cerebro-spinal fluid
- Blood vessels

Grey matter (cerebral cortex)

White matter

B The brain and spinal cord

Cerebral hemispheres

Cranial nerves

Medulla oblongata

Cerebellum

Spinal nerves

Spinal cord

Fig. 1.   Parts of the nervous system.

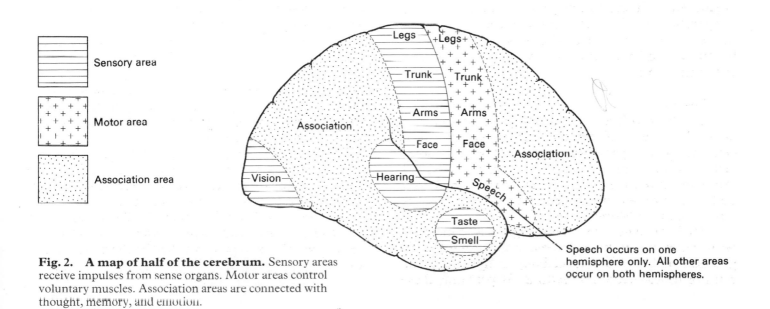

Sensory area

Motor area

Association area

Legs +Legs+

Trunk Trunk

Association

Arms Arms

Face Face

Association

Vision

Hearing

Speech

Taste

Smell

Speech occurs on one hemisphere only. All other areas occur on both hemispheres.

Fig. 2.   A map of half of the cerebrum. Sensory areas receive impulses from sense organs. Motor areas control voluntary muscles. Association areas are connected with thought, memory, and emotion.

If your hand touches something very hot you pull it away quickly without having to think about what to do. If dust blows into your eyes tears are produced and blinking occurs automatically; and if a bright light shines in your eyes the pupils contract without conscious thought. All these are examples of **reflex actions**. In technical terms a reflex action is behaviour in which a stimulus produces an immediate and automatic response.

The nerve connections responsible for a typical reflex action are illustrated in Figure 1. These connections form what is called a **reflex arc**.

Control of reflexes

Figure 2 illustrates nerve fibres which convey impulses from a reflex arc up the spinal cord to the brain and back again. These nerve fibres enable the brain to be aware of reflex actions and, in some instances, to prevent them happening if they are inconvenient. To do this, the brain sends impulses down to a reflex arc to inhibit (stop) the reflex action taking place. It does this by preventing impulses from travelling along the motor fibre of the arc. For example, provided the stimulus is not too strong, a reflex response to pain can be prevented so that it is possible to hold on to something hot even though it is burning the fingers.

Nerve connections between reflex arcs and the brain allow **conditioned reflexes** to be established.

Conditioned reflexes

In 1902 the Russian scientist Ivan Pavlov discovered how to modify certain reflex actions of dogs. A hungry dog will respond to food by producing saliva. This is a normal reflex action. Pavlov discovered that if a bell is sounded immediately before the food is presented the dog will, after many repetitions, salivate in response to the bell alone. Salivation in response to food has been replaced by salivation in response to a bell. This is called **conditioning**. Figure 3 illustrates an example of human conditioning.

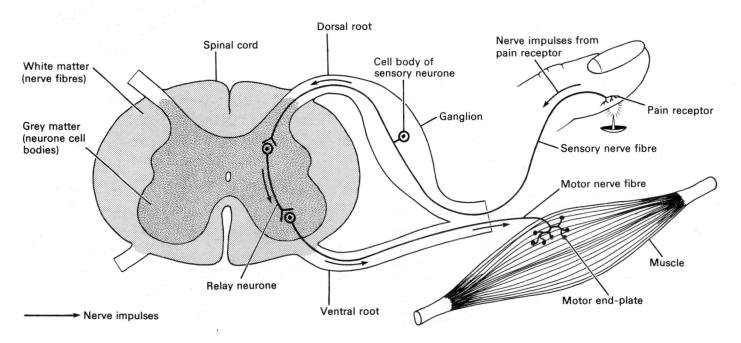

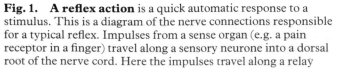

Fig. 1. **A reflex action** is a quick automatic response to a stimulus. This is a diagram of the nerve connections responsible for a typical reflex. Impulses from a sense organ (e.g. a pain receptor in a finger) travel along a sensory neurone into a dorsal root of the nerve cord. Here the impulses travel along a relay neurone and out of the nerve cord along a motor neurone to the muscle which responds to the stimulus (in this case muscles which pull the finger from the stimulus). These nerve connections are an example of a reflex arc.

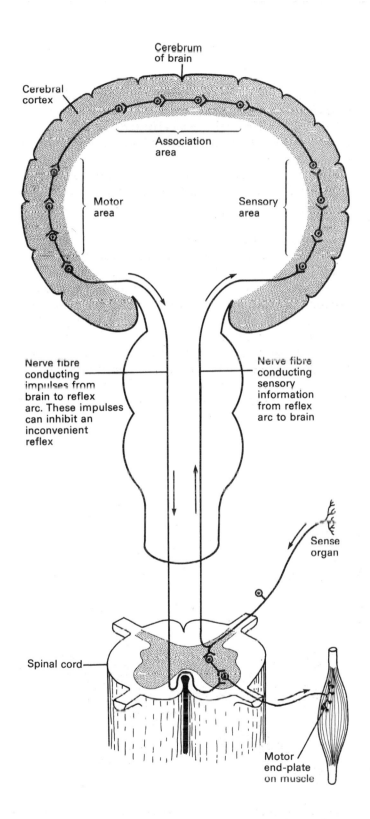

Cerebrum
of brain

Cerebral
cortex

Association
area

Motor
area

Sensory
area

Nerve fibre
conducting
impulses from
brain to reflex
arc. These impulses
can inhibit an
inconvenient
reflex

Nerve fibre
conducting
sensory
information
from reflex
arc to brain

Sense
organ

Spinal cord

Motor
end-plate
on muscle

Fig. 2.   How the brain can control reflexes. The brain is made aware of reflex actions by impulses travelling up fibres from the sensory neurone of a reflex arc. The brain can inhibit (stop) some reflexes by sending impulses down to the reflex arc.

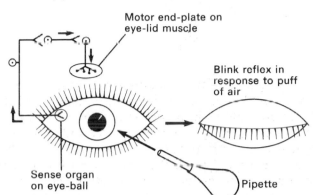

A The eye-blink reflex occurs in response to a puff of air from a pipette

Motor end-plate on
eye-lid muscle

Blink reflex in
response to puff
of air

Sense organ
on eye-ball

Pipette

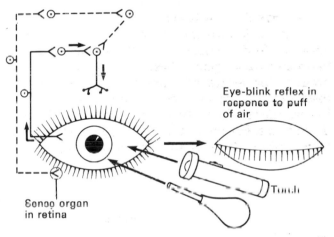

B The eye is subjected to a puff of air and to light several times. New nerve connections are established in the brain

Eye-blink reflex in
response to puff
of air

Sense organ
in retina

Touch

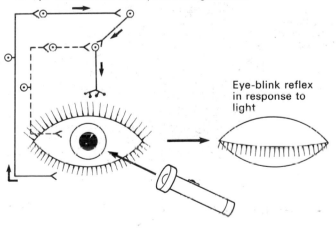

C Eye-blink occurs in response to light alone

Eye-blink reflex
in response to
light

Fig. 3.   An example of a conditioned reflex. Eyes normally blink in response to a puff of air and not light. But, after conditioning, they will blink in response to light alone.

53

Autonomic nervous system

The word autonomic means 'independent'. Consequently it is an appropriate name for the part of the nervous system which controls the unconscious internal affairs of the body; that is, those processes which are largely independent of thought.

For example, when a man runs to catch a bus his muscles are controlled by the conscious parts of his nervous system. But, at the same time, his autonomic nervous system automatically increases the rate of his breathing and heart-beat, dilates the blood vessels which carry blood to his muscles, releases sugar from his liver, and makes many other adjustments to provide for the sudden burst of activity. When the man reaches the bus and sits down his autonomic nervous system slows down all the processes which it previously speeded up. To summarize, the autonomic system regulates the body's internal environment, making automatic adjustments for activity and rest.

The autonomic nervous system consists of two parts: the **sympathetic** and the **parasympathetic**. These two systems are **antagonistic** to one another, which means they have opposite functions. The sympathetic system is mostly concerned with preparing the body for action while the parasympathetic system prepares the body for rest.

The sympathetic and parasympathetic systems are able to have opposite effects on the body because most organs and glands are penetrated by nerves from *both* systems (Fig. 1). This is called **double innervation**. When the iris of the eye, for example, receives impulses from its sympathetic nerve the pupil is dilated (enlarged), whereas parasympathetic stimulation contracts the pupil.

How sympathetic nerves prepare the body for action

In an emergency the sympathetic nervous system is able to have immediate widespread effects on the body for two reasons. First, after each sympathetic nerve fibre leaves the central nervous system it links up with twenty to thirty other fibres which transmit impulses simultaneously to many different organs and glands. Second, sympathetic nerves stimulate the adrenal glands to produce the hormone adrenalin. This hormone (described in Unit 54) is distributed quickly by the blood and stimulates organs into greater activity.

Sympathetic and adrenalin stimulation have the following effects on the body.

1. The heart-beat is accelerated and coronary blood vessles dilate, increasing blood supply to heart muscle.

2. Muscle in the bronchiole walls relaxes. This increases the diameter of these tubes so that the lungs can be inflated and deflated at greater speed.

3. Blood vessels in the skin and gut contract, reducing blood supply to these regions and diverting it to the limb muscles whose blood vessels are dilated.

4. In the liver, glycogen is converted into glucose which pours into the blood and around the body.

When the emergency is over the parasympathetic system restrains the heart-beat, contracts the bronchioles, increases blood supply to the skin and gut, and switches the liver over to converting excess blood glucose into glycogen. These and other actions adjust unconscious processes to a level suitable for leisurely conscious activity, thereby avoiding waste and excessive wear and tear on organs.

Table 3   Effects of autonomic stimulation

Sympathetic stimulation	Parasympathetic stimulation
Dilates pupil of eye	Constricts pupil
Increases heart rate	Decreases heart rate
Dilates coronary blood vessels	Constricts coronary vessels
Dilates bronchi and bronchioles	Constricts bronchi and bronchioles
Reduces peristalsis and closes gut sphincters	Increases peristalsis and opens gut sphincters
Nil	Causes production of digestive juices
Relaxes bladder wall and closes bladder sphincters	Contracts bladder wall and opens bladder sphincters
Converts glycogen to glucose in the liver	Converts glucose to glycogen in the liver
Causes production of sweat	Nil

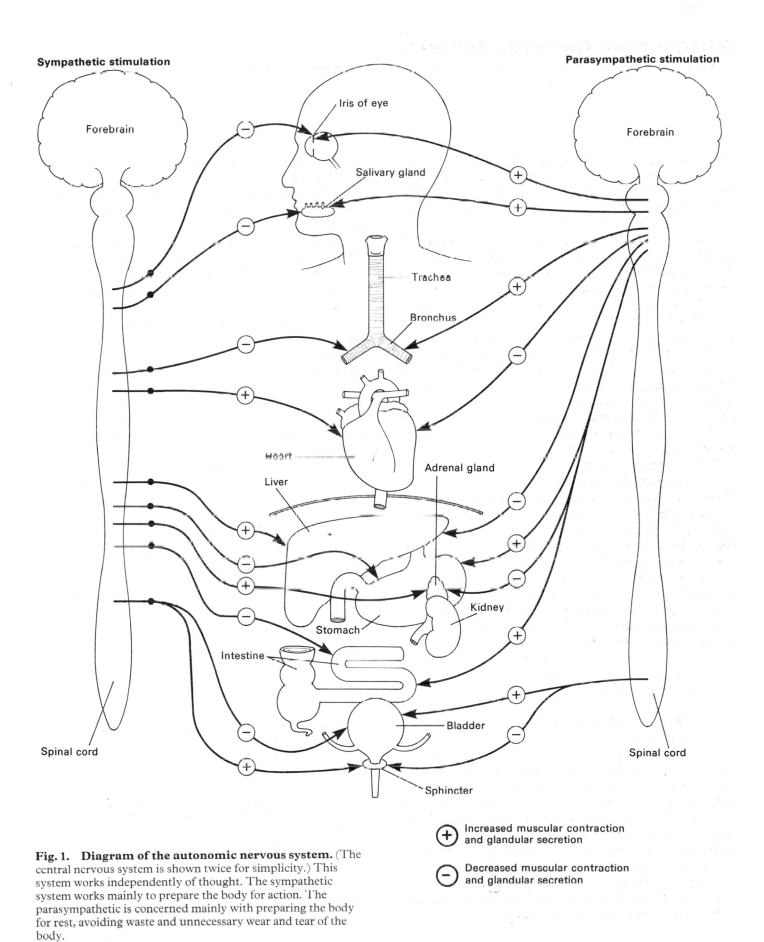

Sympathetic stimulation

Parasympathetic stimulation

Forebrain

Iris of eye

Forebrain

Salivary gland

Trachea

Bronchus

Heart

Adrenal gland

Liver

Kidney

Stomach

Intestine

Bladder

Spinal cord

Spinal cord

Sphincter

⊕ Increased muscular contraction and glandular secretion

⊖ Decreased muscular contraction and glandular secretion

Fig. 1. Diagram of the autonomic nervous system. (The central nervous system is shown twice for simplicity.) This system works independently of thought. The sympathetic system works mainly to prepare the body for action. The parasympathetic is concerned mainly with preparing the body for rest, avoiding waste and unnecessary wear and tear of the body.

54

The endocrine system

The endocrine system consists of glands which produce (secrete) chemicals called **hormones**. The human endocrine system is illustrated in Figure 2.

An endocrine gland secretes its hormones directly into the blood which transports them around the body. In contrast a non-endocrine gland, like the pancreas, secretes its products through a tube called a **duct** which leads to the area where these products are used. Since endocrine glands have no ducts they are called **ductless glands**. After entering the blood a hormone travels to all parts of the body but only parts called **target organs** respond to them.

Hormones are produced in minute quantities but their effects on the body are often profound. Some of these effects are temporary. Adrenalin for example quickens the heart-beat, releases sugar from the liver, etc., but these changes last only a few minutes. Some hormones from the pituitary gland, however, have long-lasting effects. They influence the size of the body and mental ability.

The endocrine and nervous systems are both concerned with controlling organs and tissues so that they function together as a co-ordinated **organism** rather than as a collection of independent parts. Control of this type is called **integration**. The endocrine and nervous systems also enable an organism to respond to temporary and long-lasting changes in its surroundings. This is called **adaptation**. There are, however, a number of fundamental differences in the way that hormones and nerve impulses integrate the body and adapt it to its surroundings.

Differences between hormonal and nervous co-ordination

Nerve impulses travel at great speed along nerve fibres whereas hormones travel more slowly in the bloodstream. Nerve impulses travel along a nerve to only one part of the body but hormones are transported to all parts of the body and affect several target organs. When nerve impulses reach a muscle, for example, they produce an immediate response lasting only a few seconds. Hormones produce slower responses which can last for years.

Control of endocrine glands

A few endocrine glands control themselves, but the majority are controlled by the pituitary gland. Therefore the pituitary is called the 'master gland' of the endocrine system.

A good example of how the pituitary controls other endocrine glands is the way that it regulates the flow of the hormone thyroxine from the thyroid gland. The pituitary gland is very sensitive to the amount of thyroxine in the blood. Whenever thyroxine falls below a certain level the pituitary produces a substance called **thyroid-stimulating hormone (TSH)**. This stimulates the thyroid to produce more thyroxine. When thyroxine rises above a certain level the pituitary produces less TSH and so less thyroxine is produced by the thyroid. Constant repetition of these events controls thyroxine production.

This mechanism is an example of **negative feedback**: any change in the level of thyroxine produces an opposite (negative) response which reverses the change. Information about changes in the thyroxine level is constantly 'fed back' to the pituitary so that it can make an appropriate response (Fig. 1). The pituitary controls several other endocrine glands in the same way.

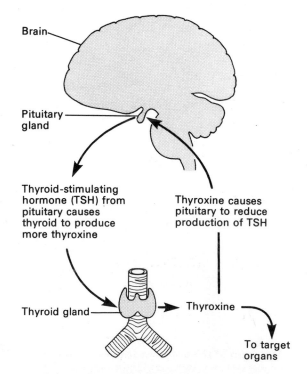

Fig. 1. The pituitary gland controls most other endocrine glands. For example, whenever thyroxine production decreases, the pituitary releases thyroid-stimulating hormone (TSH); but when thyroxine production increases the pituitary produces less TSH.

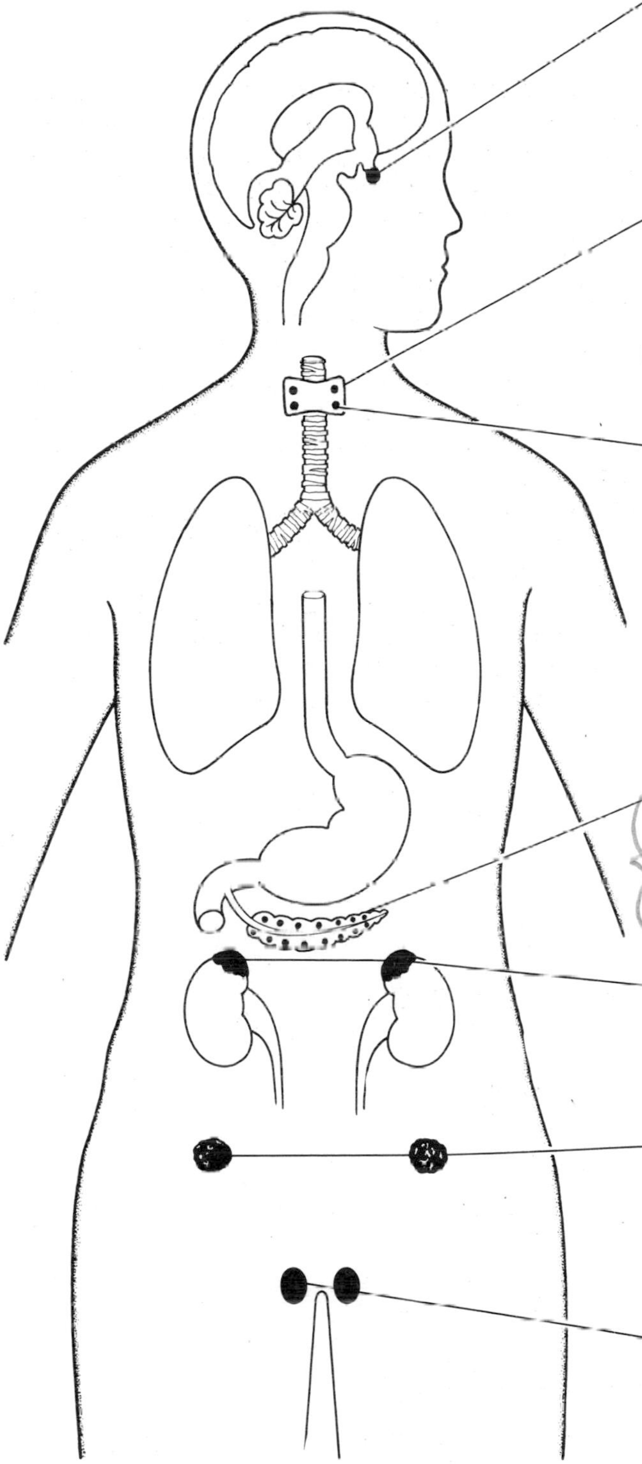

Pituitary Situated at the base of the brain. Produces hormones which control growth. Too large an amount of these hormones causes **giantism;** too little causes **dwarfism.** Other hormones produced by the pituitary cause ovaries to release eggs; testes to produce sperms; the uterus to contract and expel the foetus at birth; and the mammary glands (breasts) to produce milk. Other hormones control the amount of water in urine; and the activities of the other endocrine glands

Thyroid gland Situated in the neck in front of the wind-pipe. Produces the hormone **thyroxine,** which has a major influence on physical and mental development after birth by controlling the rate of chemical reactions in all body cells. If too little thyroxine is produced in childhood **cretinism** develops (stunted physical and mental growth). In adults too little thyroxine causes overweight, thick skin and coarse hair, slow mental and physical reactions, and premature ageing. Too much thyroxine causes underweight, restlessness, and mental instability

Parathyroid glands Four small glands embedded in the thyroid. Parathyroids are sensitive to the level of calcium in the blood. If there is too little calcium the parathyroids release the hormone **parathormone.** This increases calcium absorption in the intestine, causes withdrawal of calcium from bones, and increases reabsorption of calcium by the kidney tubules

Pancreas Situated below the stomach. One part produces digestive juice; other parts produce the hormone **insulin,** which decreases the rate at which the liver releases glucose into the blood; enables cells to absorb glucose; and stimulates the body to change glucose into fat. If too little insulin is produced the liver releases too much glucose. This causes the disease **sugar diabetes,** in which the level of glucose in the blood increases to a dangerous level

Adrenal glands Situated on top of each kidney. Produce the hormone **adrenalin.** This prepares the body for action by raising blood pressure; increasing heart-beat and breathing rates; increasing the amount of glucose sugar released from the liver; and increasing the supply of blood to the muscles and reducing the supply to the gut

Ovaries (females only) Situated in the lower abdomen, below the kidneys. The parts of the female reproductive system that produce eggs. The ovaries also produce a sex hormone called **oestrogen,** which controls development of female secondary sexual characteristics (breasts, soft skin, feminine voice); prepares the uterus so that it can receive a fertilized egg; stimulates the uterus to protect and nourish a developing baby

Testes (males only) Situated in the groin, in a sac called the scrotum. The parts of the male reproductive system which produce sperms. Testes also produce a sex hormone called **testosterone,** which controls the development of male secondary sexual characteristics (deeper voice, coarser skin, and more body hair than in females)

Fig. 2. Human endocrine glands and their functions.

55

Male reproductive system

Male sex cells are called **sperms** and are produced in organs called **testes**. Human testes are oval in shape. They are about 5 cm long and are contained in a protective bag called the **scrotum** (Figs. 1 and 2). Sperms are produced continuously from puberty (sexual maturity) until about seventy years of age. Puberty in males occurs between 10 and 17 years of age.

The inside of a testis is divided into about three hundred compartments each containing up to three narrow twisted tubes (Fig. 3). These tubes are lined with dividing cells which produce sperms. The hundreds of sperm-producing tubes join together and form a smaller number of **collecting ducts**. Sperms move out of the testis through these ducts into a single coiled tube called the **epididymis**. This is a storage area for sperms. Each epididymis leads into a sperm duct, or **vas deferens**. The sperm ducts, one from each testis, rise up the body until they are joined by a duct from a **seminal vesicle**. The sperm ducts join together near the base of the bladder forming a single tube called the **urethra**. This junction occurs inside the **prostate gland**. Lower down, the urethra is joined by a duct from Cowper's gland and then it leads to the outside of the body through the **penis**.

The walls of the penis contain sponge-like spaces called **erectile tissue**. When a male is sexually stimulated the erectile tissue fills with blood making the penis erect and firm. During sexual intercourse the erect penis is inserted into the vagina of the female and moved back and forth. These movements stimulate sense organs in the penis and eventually cause rhythmic contractions in the walls of the urethra which propel a liquid called **semen** into the female. This event is called an **ejaculation**. Semen consists of millions of sperms floating in a liquid from the seminal vesicles and from the prostate and Cowper's glands. This liquid nourishes the sperms and causes their tails to make the swimming movements which propel them through the female reproductive system to an egg.

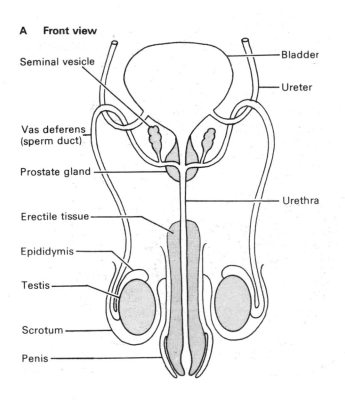

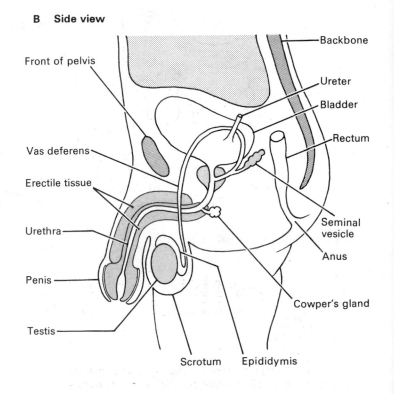

Fig. 1. Diagram of male reproductive system (front view with testes drawn further apart than normal).

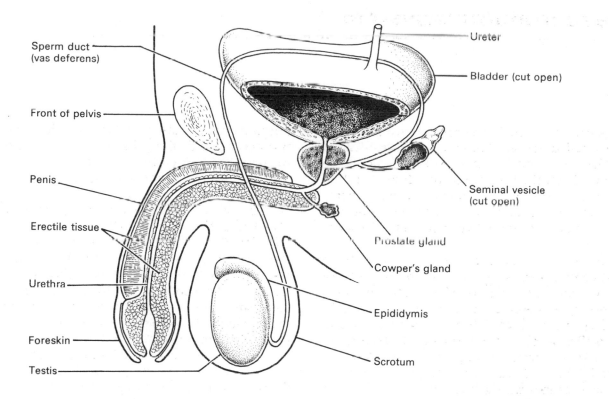

Labels for Fig. 2 (clockwise from top left):
- Sperm duct (vas deferens)
- Front of pelvis
- Penis
- Erectile tissue
- Urethra
- Foreskin
- Testis
- Ureter
- Bladder (cut open)
- Seminal vesicle (cut open)
- Prostate gland
- Cowper's gland
- Epididymis
- Scrotum

Fig. 2. Male reproductive system (side view). During sexual intercourse the penis becomes erect and is inserted into the vagina of the female. Rhythmic contractions of the sperm ducts propel semen into the female. Semen consists of sperms and liquid from the seminal vesicles, the prostate gland, and Cowper's gland.

Cross-section through a sperm-producing tube

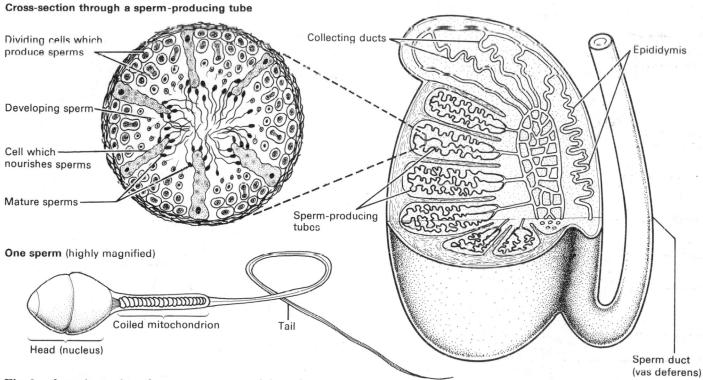

Labels for Fig. 3:
- Dividing cells which produce sperms
- Developing sperm
- Cell which nourishes sperms
- Mature sperms
- Collecting ducts
- Epididymis
- Sperm-producing tubes
- Sperm duct (vas deferens)

One sperm (highly magnified)
- Head (nucleus)
- Coiled mitochondrion
- Tail

Fig. 3. A testis consists of compartments containing tubes lined with dividing cells which make sperms. Sperms are sex cells. The head of a sperm contains the cell nucleus and this is propelled by the tail.

111

Female sex cells called **ova** (*singular*: ovum) are produced by two **ovaries**. Human ovaries are 3 cm long and are attached to the back wall of the abdomen above the kidneys (Fig. 1). Ova begin forming in the ovaries before birth, and a newborn baby girl has a million or more partly formed ova in her ovaries. No new ova develop from this time onwards, however, and during childhood most existing ova disintegrate.

Puberty (sexual maturity) occurs in girls between eight and fifteen years of age, and by this time each ovum is encased in a mass of cells called a **follicle**. Once a month the pituitary gland secretes a hormone which causes one of these follicles to absorb liquid, expand, and burst, releasing its ovum from the ovary. This event is called **ovulation** (Figs. 2 and 3). Ovulation occurs once a month, unless interrupted by pregnancy, until about the age of fifty.

After leaving an ovary the ovum passes into a **fallopian tube**. Each fallopian tube has a funnel-shaped opening and is lined with microscopic hairs called cilia (Fig. 2). Peristaltic contractions in the walls of the fallopian tubes sweep an ovum along until it reaches a wider thicker-walled tube called the **womb**, or **uterus**. An ovum which has been fertilized by a sperm develops into a baby in the uterus.

A ring of muscle called the **cervix** closes the lower end of the uterus where it joins another tube, the **vagina**. The vagina extends to the outside of the body where there is an opening called the **vulva**.

A Front view

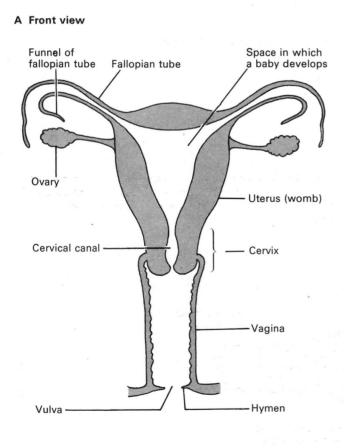

B Side view

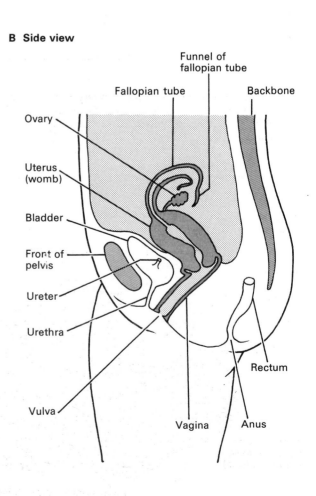

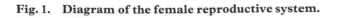

Fig. 1. Diagram of the female reproductive system.

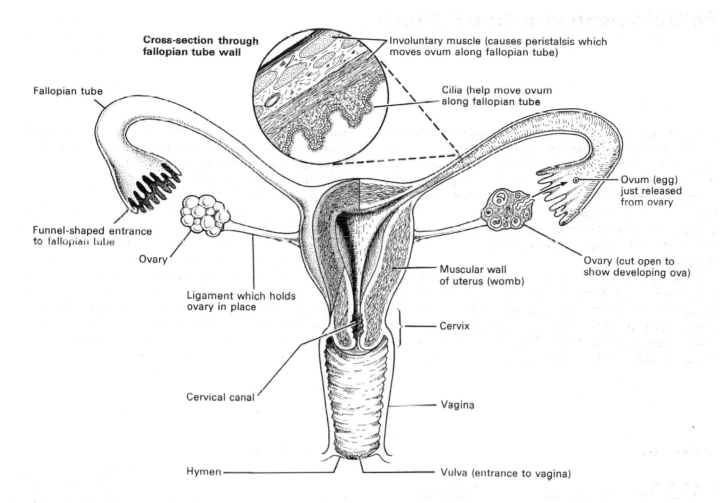

Cross-section through fallopian tube wall

Fallopian tube

Involuntary muscle (causes peristalsis which moves ovum along fallopian tube)

Cilia (help move ovum along fallopian tube

Funnel-shaped entrance to fallopian tube

Ovary

Ligament which holds ovary in place

Cervical canal

Hymen

Ovum (egg) just released from ovary

Ovary (cut open to show developing ova)

Muscular wall of uterus (womb)

Cervix

Vagina

Vulva (entrance to vagina)

Fig. 2. Female reproductive system (left side cut open). Once a month an ovary releases an ovum. It is moved along a fallopian tube by peristaltic contractions and by the action of cilia.

Fig. 3.   Inside an ovary each ovum is enclosed in a cluster of cells called a follicle. One at a time the follicles absorb water, expand, burst open, and release an ovum. This is called ovulation.

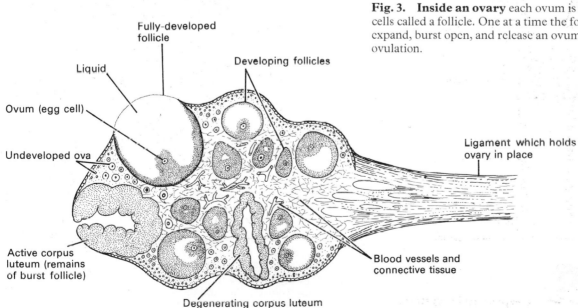

Fully-developed follicle

Developing follicles

Liquid

Ovum (egg cell)

Undeveloped ova

Ligament which holds ovary in place

Active corpus luteum (remains of burst follicle)

Blood vessels and connective tissue

Degenerating corpus luteum

57

The menstrual cycle

Starting at puberty the female reproductive system undergoes a regular sequence of events, repeated monthly, called the **menstrual cycle**. One menstrual cycle lasts about twenty-eight days, although it can be longer or shorter than this.

During one menstrual cycle an ovum is released from an ovary and the lining of the womb (uterus) is renewed. It develops to a stage at which it can receive and nourish the ovum should this be fertilized and begin to grow into a baby. These events are controlled by hormones from the ovaries and from the pituitary gland, and are summarized in Figure 1.

Events during one menstrual cycle

1. From day one to about day five of a cycle the lining of the uterus breaks down into loose tissues which flow out of the vagina together with 50 to 250 cm³ of blood. The technical term for this process is **menstruation**, which is commonly known as 'a period'.

2. When bleeding stops an ovary is stimulated by secretion of a chemical called **follicle-stimulating hormone** from the pituitary gland. The ovary responds to this hormone by growth of one or more follicles. As it grows in size the follicle secretes increasing amounts of a hormone called **oestrogen**. The main effect of oestrogen is to stimulate growth of a new lining in the uterus. This lining consists of a mass of blood vessels and glandular tissue. Oestrogen also stimulates the breasts (mammary glands) to grow the small tubes (ducts) which could eventually carry milk to the nipples.

3. The rising level of oestrogen in the blood acts as a feed-back mechanism: it causes the pituitary to stop producing follicle-stimulating hormone and begin producing another hormone which brings about ovulation. This usually occurs sometime between days 13 and 15 of the cycle, but it can happen earlier or later than this. Immediately after ovulation a woman is said to be **fertile**, and can **conceive**—become pregnant—if she has sexual intercourse. Since it is difficult to know exactly when ovulation takes place, and since sperms can live for up to three days in the uterus after intercourse, most women must assume that they are fertile from at least day 11 to day 17 of their cycle (Fig. 1).

4. The part of a follicle left behind in the ovary after ovulation forms a large yellow object called a **corpus luteum** (Fig. 56.3). This acts as a temporary endocrine organ and produces the hormone **progesterone**. Progesterone stimulates further growth of the uterus wall, and it stimulates the breasts to begin developing the glandular tissues which make milk.

5. Meanwhile the ovum has been sucked into a fallopian tube and is moving towards the uterus. If it is not fertilized within thirty-six hours after ovulation the ovum dies.

6. Death of the ovum is soon followed by a gradual reduction in size of the corpus luteum and a reduction in the amount of progesterone it produces. Lack of progesterone brings about another period about 14 days after ovulation.

Menopause

Men can continue producing sperm throughout life but women lose their ability to have children between forty-two and fifty-five. **Menopause** is the technical name for this loss of fertility, but it is commonly known as 'change of life'. At this time ovulation and menstruation stop and the reproductive organs decrease in size.

Fig. 1 (opposite). The menstrual cycle is a sequence of events, repeated monthly, in the female reproductive system. These events are controlled by hormones from an ovary and from the pituitary gland. During one 28 day cycle an ovum is released from an ovary and the womb is prepared to receive this ovum should it be fertilized and develop into a baby.

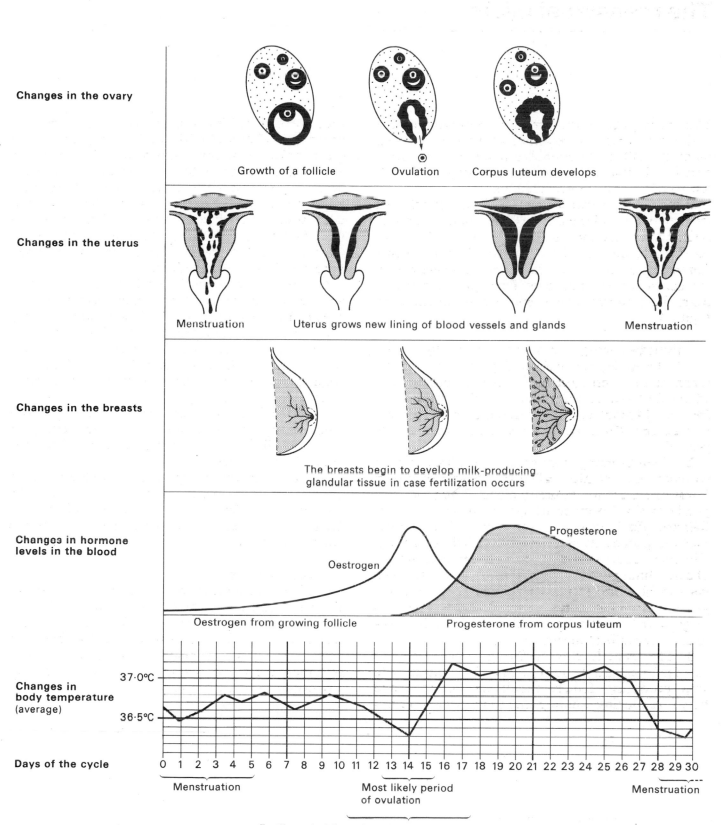

Changes in the ovary

Growth of a follicle   Ovulation   Corpus luteum develops

Changes in the uterus

Menstruation   Uterus grows new lining of blood vessels and glands   Menstruation

Changes in the breasts

The breasts begin to develop milk-producing
glandular tissue in case fertilization occurs

**Changes in hormone
levels in the blood**

Progesterone

Oestrogen

Oestrogen from growing follicle   Progesterone from corpus luteum

**Changes in
body temperature**
(average)

37·0°C

36·5°C

Days of the cycle

0 1 2 3 4 5 6 7 8 9 10 11 12 13 14 15 16 17 18 19 20 21 22 23 24 25 26 27 28 29 30

Menstruation   Most likely period
of ovulation   Menstruation

Fertile period (based on the assumption that sperms
can live for three days in the uterus, that ovulation usually
occurs between days 13 and 15, and that an ovum can live
for 36 hours after ovulation)

115

58

Fertilization and implantation

Fertilization

About 5 cm³ of semen, containing up to 300 million sperms, are ejaculated into the female reproductive system during copulation, but only one sperm is needed to fertilize an ovum. A large number of sperms are necessary because many fail to complete the long journey to the area where fertilization occurs.

Sperms swim through the cervix into the uterus and up the fallopian tubes. Usually sperms are sucked into the uterus by muscular contractions of its walls caused by the stimulation of sexual intercourse.

If a ripe ovum is present at the time of copulation fertilization occurs, usually in the upper third of a fallopian tube. The ovum is surrounded by a membrane which a sperm must penetrate. The first sperm to reach the ovum releases a chemical which softens this membrane at the point of contact, allowing the sperm to enter and move into the ovum's cytoplasm. Immediately after penetration by a sperm the membrane undergoes rapid changes which make it thicker and cause it to lift away from the surface of the ovum. It is now called the **fertilization membrane**, and forms a barrier to the entry of other sperms (Fig. 1).

Only the head of a sperm enters the ovum. The head moves through the cytoplasm towards the nucleus of the ovum and the two fuse together. This fusion brings together the hereditary characteristics of the mother and father.

A fertilized ovum is called a **zygote**. It is propelled towards the uterus mainly by wave-like contractions (peristalsis) of a fallopian tube, helped by movements of the cilia lining the tube. During this journey, which takes about 7 days, the zygote begins to divide (Fig. 2). By the time it reaches the uterus it consists of a hundred or more cells which form a hollow ball called an **embryo**. The embryo now establishes contact with the uterus wall which has been prepared to nourish it. This process is called implantation.

Implantation

There is a theory that the embryo produces digestive enzymes which digest a hole in the uterus lining into which it sinks. It is possible, however, that the uterus lining dissolves away automatically as the embryo comes into contact with it. Whatever the mechanism the embryo is soon firmly embedded among the glands and blood vessels lining the uterus (Fig. 2).

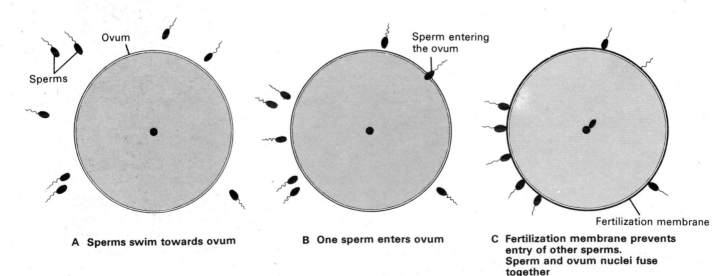

A **Sperms swim towards ovum**   B **One sperm enters ovum**   C **Fertilization membrane prevents entry of other sperms. Sperm and ovum nuclei fuse together**

Fig. 1. Fertilization is the fusion of one sperm with an ovum. This event brings together the hereditary characteristics of the mother and father.

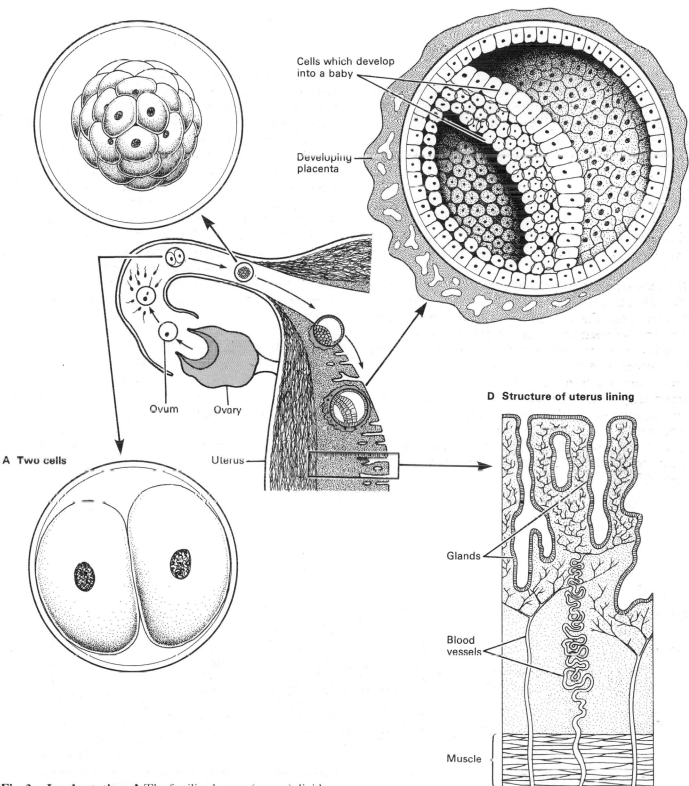

B Hollow ball of cells

C Implanted embryo with developing placenta

Cells which develop into a baby

Developing placenta

Ovum   Ovary

D Structure of uterus lining

Uterus

Glands

Blood vessels

Muscle

A Two cells

Fig. 2. Implantation. A The fertilized ovum (zygote) divides into two cells. **B** Cell division continues to make a ball of cells. **C** The ball becomes hollow and is called an embryo. The embryo sinks into the lining of the uterus. This is called implantation. **D** Structure of the uterus lining.

59

Pregnancy and the placenta

The period of development between fertilization and birth is called **pregnancy**. Human pregnancy lasts about nine months. During this time hormones in the mother's blood prevent the monthly occurrence of menstruation and ovulation. Implantation of a developing embryo is the signal which ensures that these hormones flow.

If implantation takes place, cells surrounding the embryo produce hormones which prevent breakdown of the corpus luteum so that it continues producing progesterone. This hormone stimulates further growth of the uterus and milk-producing tissues in the breasts.

A developing embryo must have oxygen and food. At first it obtains these directly from its mother's blood flowing through vessels in the lining of the uterus. Within three weeks of implantation, however, food and oxygen are absorbed within an organ called a **placenta**.

A placenta is a disc-shaped organ with millions of tiny root-like outgrowths called **villi**. These villi are embedded in the lining of the uterus (Figs. 1 and 2). A placenta is connected with the embryo by a tube called the **umbilical cord**. The embryo's heart pumps blood through blood vessels in the umbilical cord into the placenta where it flows through capillaries in the placental villi (Fig. 2). The villi hang inside spaces in the uterus wall filled with the mother's blood. In these spaces food and oxygen pass from the mother's blood into the embryo's blood, while at the same time carbon dioxide and other wastes pass from the embryo's blood into the mother's blood.

The placenta is a temporary endocrine organ. It produces large amounts of oestrogen and progesterone. The oestrogen prevents the pituitary gland from making follicle-stimulating hormone so that follicles do not develop during pregnancy. The progesterone ensures that the uterus grows at the same rate as the baby and also that the breasts are ready to produce milk soon after the baby is born.

The baby develops inside a bag of liquid called the **amnion** (Figs. 1 and 2). The amnion acts as a shock-absorber—the liquid in it cushions the baby against jolts and knocks as the mother moves about. It also helps prevent the baby being injured if anything hits or presses against the mother's body.

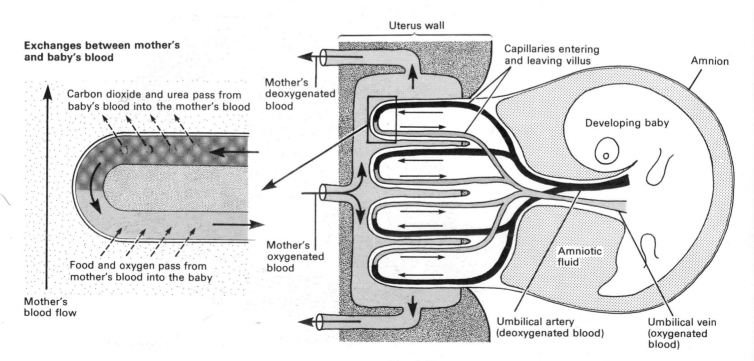

Exchanges between mother's and baby's blood

Carbon dioxide and urea pass from baby's blood into the mother's blood

Food and oxygen pass from mother's blood into the baby

Mother's blood flow

Uterus wall

Mother's deoxygenated blood

Mother's oxygenated blood

Capillaries entering and leaving villus

Amnion

Developing baby

Amniotic fluid

Umbilical artery (deoxygenated blood)

Umbilical vein (oxygenated blood)

Fig. 1. Diagram of an embryo and its placenta. The embryo's heart pumps blood through the placenta where it passes through capillaries in placental villi. The villi are in blood-filled spaces in the uterus wall. Food and oxygen pass from the mother's to the embryo's blood while carbon dioxide and urea pass from the embryo's to the mother's blood.

Structure of the placenta

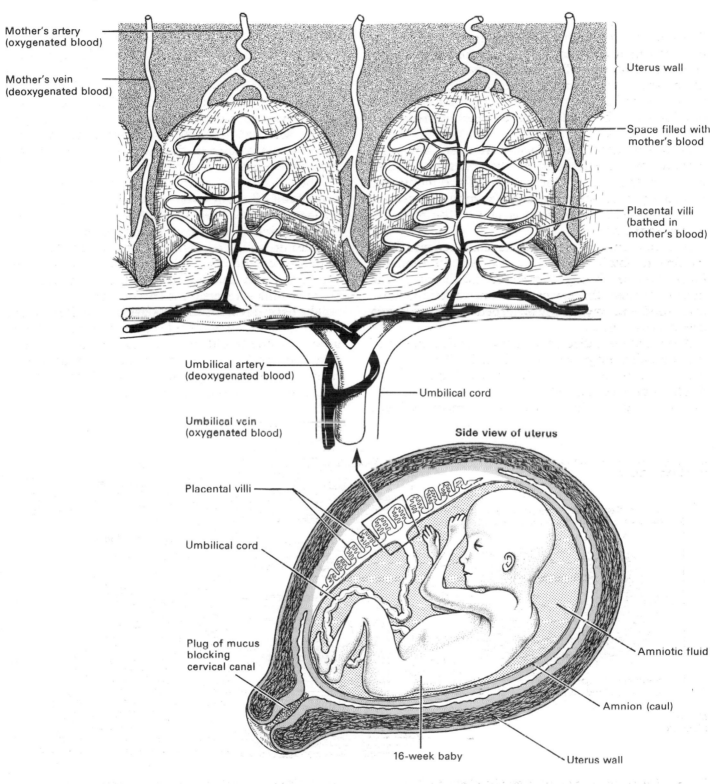

Mother's artery (oxygenated blood)

Mother's vein (deoxygenated blood)

Uterus wall

Space filled with mother's blood

Placental villi (bathed in mother's blood)

Umbilical artery (deoxygenated blood)

Umbilical cord

Umbilical vein (oxygenated blood)

Side view of uterus

Placental villi

Umbilical cord

Plug of mucus blocking cervical canal

Amniotic fluid

Amnion (caul)

16-week baby

Uterus wall

Fig. 2.   Structure of the placenta.

60

Development before birth

Ideally the age of an unborn baby should be calculated from the moment of fertilization. But this is difficult because it is not always possible to know exactly when fertilization occurred. To overcome this difficulty it is usual to calculate a developing baby's age from the first day of the mother's last period, since this date can be calculated with some certainty. This system gives the **menstrual age** of the baby, and is used in the following descriptions. It must be remembered, however, that the menstrual age is always about two weeks too much.

Week 5

A fertilized ovum is smaller than a printed full stop, but by week 5 (about three weeks after fertilization) it is about 5 mm long. It has a bulge which will soon be a head; the brain, eyes, and ears are beginning to develop; and a simple heart pumps blood through a small placenta. The embryo has a long tail and small lumps on its sides are forming into arms and legs.

Weeks 8 to 10

During this period the embryo becomes clearly human in appearance and from now until birth is called a **foetus**. It has a face, and limbs with fingers and toes. Most of its internal organs are fully formed and will continue to grow until birth.

Week 20

The foetus is now about 25 cm long. The mother can feel it moving occasionally in her womb and its heart-beats can be heard with a stethoscope. Sex organs are developed to a stage at which differences between a boy and girl are obvious.

Week 28

By this time the foetus is 35 to 38 cm long. It has soft downy hair and eye-lashes, and can open its eyes. Finger and toe nails have hardened and milk teeth are developing in its jaws. Its body is covered with white grease called **vernix** which stops the skin becoming waterlogged with amniotic fluid. Its organs have now grown to a stage at which the baby is just able to survive with expert care if removed from its mother. In technical terms it is **legally viable**.

Week 40

The pregnancy is now **at term**—the baby is ready to be born. It weighs between 3 and 3.5 kg and is about 50 cm long.

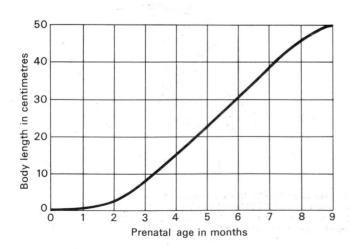
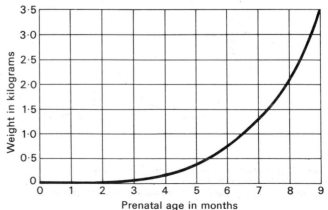

Fig. 1. Changes in the length and weight of a baby up to birth (prenatal period).

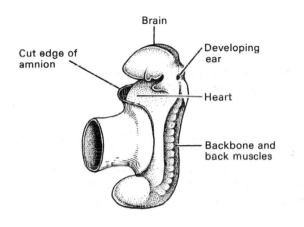

Brain

Cut edge of amnion

Developing ear

Heart

Backbone and back muscles

⊥ Actual size

Week 3 The embryo is about 2·0 mm long. Already the developing brain, ears, heart, and backbone can be seen

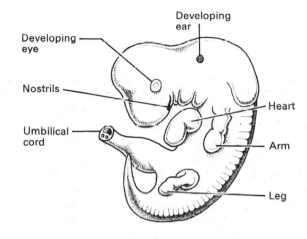

Developing eye

Developing ear

Nostrils

Umbilical cord

Heart

Arm

Leg

⊥Actual size

Weeks 4 to 5 The embryo is about 5 mm long. Its brain, eyes, and ears are developing and a heart pumps blood to a placenta. Limbs are represented by lumps on each side of the body

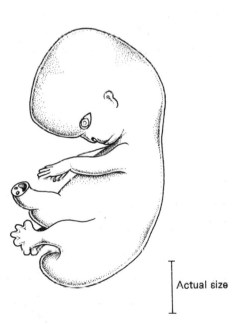

Actual size

Weeks 6 to 7 The embryo is about 14 mm long. Its limbs well-formed. Nose and mouth are developing and lungs, liver, and kidneys are taking shape

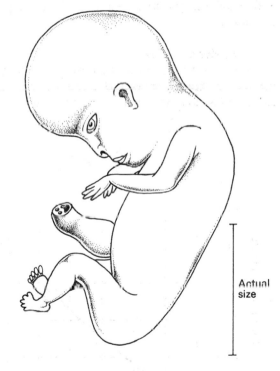

Actual size

Weeks 8 to 9 The body is about 35 mm long. It is now called a foetus and is human in appearance. The body is almost completely formed and will continue growth until birth

Fig. 2. Development of a baby inside the uterus.

During the months before birth the uterus walls develop a thick layer of muscle. Birth begins when hormones in the mother's blood cause this muscle to start contracting rhythmically. At first the contractions are weak and far apart but gradually become stronger and closer together and cause **labour pains**. Eventually the contractions push the baby out of the mother's body.

A few weeks before birth most babies turn within the uterus until their head points downwards towards the cervix (Fig. 2A). Birth occurs in three stages. First the cervix expands allowing the baby's head to pass into the vagina (Fig. 2B). Second, contractions push the baby through the vagina and out of the mother's body. This is called **delivery** (Fig. 2C). Third, further contractions push the placenta, or **after-birth**, out of the mother's body (Fig. 2D).

After leaving the mother a baby takes its first breath and usually begins crying. Crying is a reflex action which establishes regular breathing. These events bring about changes in the baby's circulatory system.

Circulatory changes at birth

A foetus obtains oxygen from its placenta. Its lungs do not function and receive very little blood. But at birth the placenta stops functioning and its blood supply is cut off. There is then a rapid alteration in circulation so that blood is diverted to the lungs as they inflate with air for the first time.

Figure 1 shows how this happens. In a foetus blood entering the right atrium is not pumped to the lungs; it by-passes them in two ways. Some blood passes from the right atrium to the left atrium through a hole called the **foramen ovale**, and blood which enters the pulmonary artery is transferred to the aorta by a vessel called the **ductus arteriosus**.

At birth blood supply to the placenta is cut off and some blood begins flowing to the lungs. When this blood returns to the heart it raises blood pressure in the left atrium which presses a flap of tissue over the foramen ovale sealing it up so that blood no longer crosses from the right to the left atrium (Fig. 1). A few hours later the ductus arteriosus closes making all the blood pumped from the right ventricle flow to the lungs.

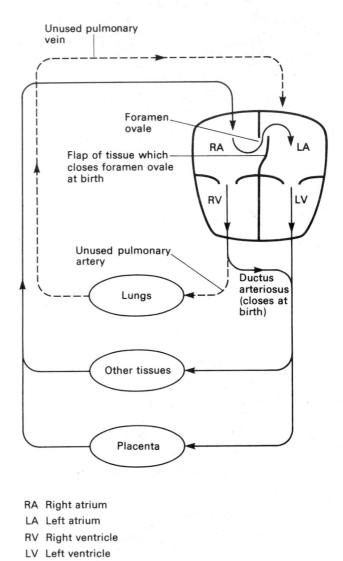

RA Right atrium
LA Left atrium
RV Right ventricle
LV Left ventricle

Fig. 1. Changes in circulation at birth. In a foetus blood entering the right atrium is diverted from the lungs in two ways. Some blood passes through the foramen ovale into the left atrium, and the remainder is diverted from the pulmonary artery through the ductus arteriosus into the aorta. At birth the foramen ovale and ductus arteriosus close up so that blood is pumped from the right ventricle to the lungs.

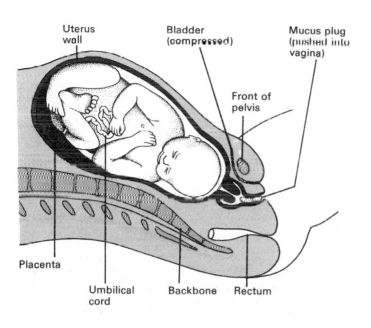

Uterus wall

Bladder (compressed)

Mucus plug (pushed into vagina)

Front of pelvis

Placenta

Umbilical cord

Backbone

Rectum

A A few days before birth the baby turns in the uterus until its head points towards the cervix. Birth begins when the uterus starts to contract rhythmically

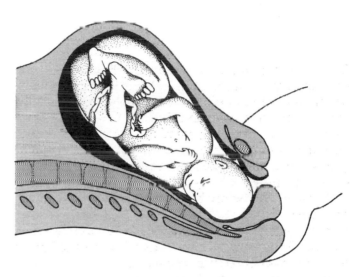

B During the first stage of birth contractions force the head onto the cervix stretching it until the head passes through to the vagina. The mucus plug is displaced and the amnion usually bursts

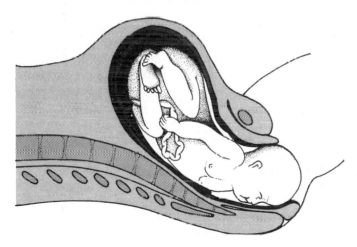

C During the second stage of birth contractions are strong and close together. The baby moves to a face-down position and is pushed out of the mother's body

Placenta becoming detached from uterus wall

Umbilical cord

D During the third stage of birth further contractions of the uterus push the placenta and umbilical cord (after-birth) out of the mother's body

Fig. 2. The birth of a baby.

62

The newborn baby

A newborn baby displays several reflex responses to changes in its surroundings (Fig. 1). Many of these reflexes are used to check the baby's health and the condition of its nervous system.

Feeding

Both breast feeding and bottle feeding produce normal healthy babies.

Breast feeding Most mothers are able to breast feed their babies, and if they can this method has a number of advantages over bottle feeding. Breast milk costs nothing and does not have to be prepared. It is available immediately, which is convenient, especially for the 2 a.m. feeds. Breast milk is always fresh and, provided the nipples are clean, free from germs. It contains ideal proportions of all the foods a baby requires and is digested rapidly and easily. There is evidence that, compared with bottle-fed babies, those which are breast-fed are less likely to have digestive upsets, skin disorders (including nappy rash), and respiratory infections.

Babies are usually put to the breast within eighteen hours of birth. Some hospitals, however, believe that this should be done as soon as possible after delivery because the sucking of a baby at a breast produces a reflex response in the mother which causes her pituitary gland to release the hormones **prolactin** and **oxytocin**. Prolactin stimulates milk production in the breasts and oxytocin causes the breasts to eject milk. Oxytocin also causes the uterus to contract to its former size.

When a baby is first put to the breast it receives a thick liquid called **colostrum**, which is particularly rich in proteins. True milk does not appear until two or three days after birth but during this time a baby has little appetite, sleeps most of the time, and usually loses some weight. By the time true milk appears most babies have become more wakeful and hungry. The baby sucking at the breast stimulates production of milk so that the breasts adjust to the baby's demands, gradually increasing milk production as the baby grows and wants more.

Bottle feeding All artificial or formula milks for use in bottle feeding are based on cows' milk with various sugars and other substances added to make them more like human milk. Table 4 gives the differences between cow and human milk. The main advantages of bottle feeding are that the baby's food intake can be measured and that people other than the mother can help with feeding. The main disadvantages of bottle feeding are that formula milk is expensive and, although it is a good imitation of breast milk, it is not so easily digested. Furthermore, unless carefully cleaned and sterilized a bottle can transmit germs to a baby.

Care of mother and baby

Mother Soon after delivery a mother should carry out regular exercises to strengthen the muscles stretched out of shape during birth.

The large wound in the uterus wall where the placenta was attached takes four to six weeks to heal and during this time a substance called **lochia** is discharged from the wound and passes out of the vulva. The vulva must be kept extremely clean because, until the uterus heals, the reproductive and urinary passages can easily become infected with germs. Regular bathing is essential, preferably in water to which a little salt has been added.

Baby It takes a few weeks before a baby's temperature-regulating mechanism begins to function efficiently and before it builds up a layer of fat beneath its skin. At first, therefore, a baby's room should be heated to about 20 °C. The air in a heated house tends to be very dry and this can damage the delicate membranes lining a baby's respiratory passages. So it is beneficial to take a baby out of doors in warm clothing for an hour or two a day.

A baby should be bathed once a day, preferably before a meal. The mother should wash her hands before and after feeding her baby. If bottles are used they should be cleaned in water and detergent with a bottle brush and placed in sterilizing liquid. Teats should be boiled after use.

Table 4   Human and cows' milk compared

	Human g per 100g	Cow g per 100g
Water	88.0	88.0
Protein	1.2	3.3
Fat	3.5	3.6
Lactose	6.5	4.7
Vitamin C	4 to 8 mg	0.7 to 3 mg

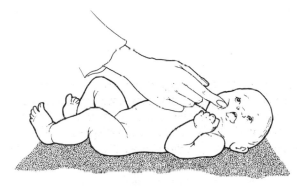

A Rooting reflex If either cheek is touched, the baby turns its head in the direction of the touch. This enables it to find the nipple of its mother's breast and, together with the sucking reflex, enables it to obtain food.

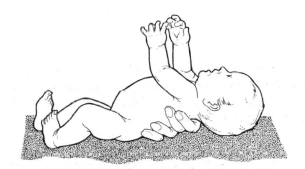

D Moro or startle reflex When startled a baby throws out its arms and legs, then pulls them back with the fingers curved, as if to catch hold of something.

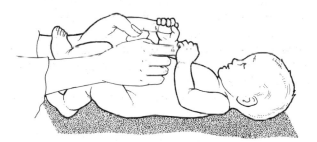

B Grasp reflex A baby will automatically grasp an object placed in its palm.

E Resting position When resting on its back, a newborn baby lies with its head to one side, with the arm and leg on that side extended and the opposite arm and leg bent.

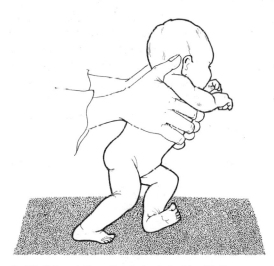

C Walking reflex If a baby is held with its feet touching the ground and moved forwards it will perform walking movements.

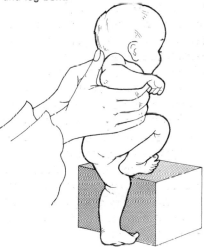

F Stepping reflex If a baby is held with the front of one leg in contact with an object, it will raise its other leg and step onto the object.

Fig. 1. Primitive reflexes in a newborn baby. These are used to check a baby's health and the condition of its nervous system.

Within three months the primitive reflexes have been replaced by learned responses.

63

Development of co-ordination

The rate at which a baby learns to co-ordinate its muscles depends mostly upon the rate at which its nervous system is maturing. Millions of new nerve connections have to be made and countless nerve fibres have to form their insulating sheaths of myelin, described in Unit 9, before impulses can pass freely between the nervous system and all the muscles of the body.

The appearance of co-ordinated movements is called **motor development**. In the first four years of life this includes the appearance of skills such as holding the head erect, sitting, crawling, rolling over, standing, walking, running, and manipulating objects with the hands (Figs. 1 and 3).

Motor development proceeds at a different rate in every child. The ages quoted in this Unit are no more than a rough guide. Normal variation can result in much faster, and much slower, rates of development.

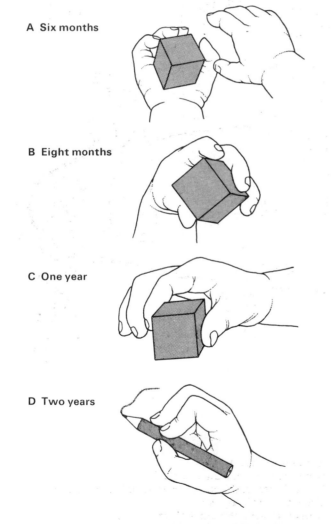

A Six months

B Eight months

C One year

D Two years

Fig. 1. Development of manipulation. A By six months a child can reach out, grasp objects, and pass them from hand to hand. **B** By eight months objects are held between the thumb and opposite fingers. **C** By one year small objects can be picked up using the thumb and forefinger. **D** By two years a child can unscrew things, and soon after a pencil is held like an adult.

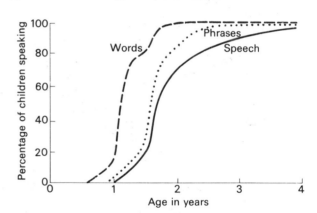

A Appearance of words, phrases, and speech

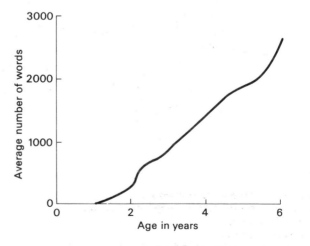

B Number of words understood

Fig. 2. Development of speech. Some vowel sounds are spoken by seven weeks and the first consonants are spoken by twenty weeks. The first words are usually spoken just before one year and by two years phrases are spoken. By three years most children can speak fluently.

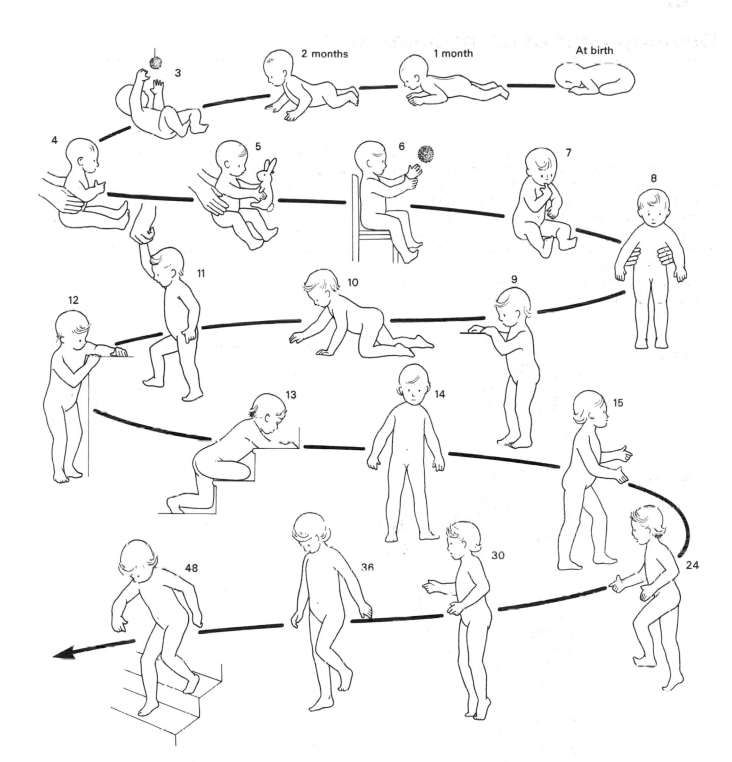

Fig. 3. Motor development.

At birth primitive reflexes only (see Unit 63).
1 month chin up,
2 months chest up,
3 months reaches for but misses objects,
4 months sits with support,
5 months grasps objects,
6 months can sit on chair, reach for and catch objects,
7 months sits unsupported,
8 months stands with some support,
9 months stands alone but holding furniture,

10 months crawls rapidly,
11 months walks holding one hand,
12 months uses furniture to rise to standing position,
13 months crawls up stairs,
14 months stands alone and unsupported,
15 months walks alone,
24 months runs, and can pick up things from standing position without over balancing,
30 months stands on tiptoe and jumps with both feet,
36 months balances on one foot,
48 months walks downstairs placing only one foot on each step.

64

Growth, development, and ageing

Every child's pattern of growth and development is different from every other child's, because it depends upon many different conditions.

Development is affected by things which happen even before birth. A mother's hereditary and physical characteristics affect her child and so do her health, her diet, and whether she smokes, or drinks alcohol. The child's own hereditary characteristics affect its development in many ways, some of which are not yet understood, and equally important are a balanced diet and plenty of exercise.

Height and weight increase at different rates (Fig. 1). Different parts of the body also grow at different rates. The nervous system and skull have almost reached adult size by six years of age, whereas the reproductive organs do not grow rapidly until ten or twelve years (Fig. 2).

Unequal growth of the body causes its proportions to change with age. A baby's head, for instance, makes up a quarter of its height. In adults the head makes up only an eighth of the height (Fig. 3).

Growth occurs in a series of 'spurts' (Fig. 4). A period of rapid growth in height is followed by a 'filling-out' of the body with little height increase. The first growth spurt occurs in the first year, followed by another between five and seven years. The final growth spurt starts with puberty.

Puberty generally occurs earlier in girls than boys and transforms the body of a child into that of an adult. These changes are brought about by the hormone **testosterone** in boys and **oestrogen** in girls. In addition to speeding up growth of reproductive organs, these hormones are also responsible for development of the **secondary sexual characteristics**: rounded hips and breasts in females, compared with greater muscular development, coarser body hair, and a deeper voice in males.

Ageing, or **senescence**, occurs very slowly at first but speeds up later in life. Muscle power, for example, is at its maximum around twenty years of age and is about the same at forty, but declines after seventy. The skin becomes thin, wrinkled, and less elastic with age. Fractured bones mend more slowly, kidneys work less efficiently, and loss of elasticity in the arteries reduces efficiency in the circulatory system. Old people should not eat excessive amounts of fat or carbohydrate since obesity reduces life expectancy, but an adequate supply of vitamins is essential for a healthy old age.

A Boys

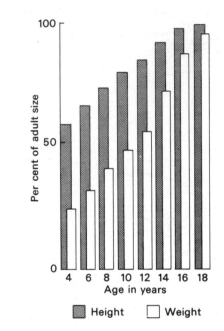

B Girls

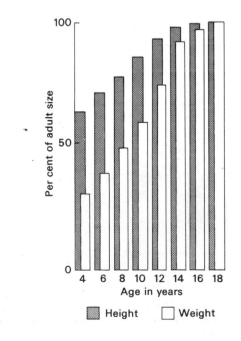

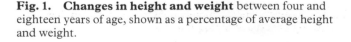

Fig. 1. Changes in height and weight between four and eighteen years of age, shown as a percentage of average height and weight.

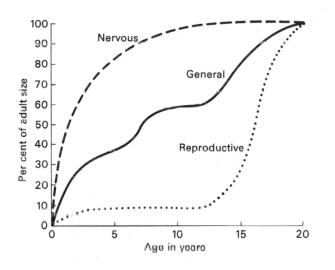

Fig. 2. **Growth of body parts.** Different parts of the body grow at different rates. The body in general (muscles, skeleton, lungs, blood volume, and intestines) grows in spurts: in infancy (first year), childhood (5–7 years), and towards puberty (11–16 years). The nervous system (brain, spinal cord, eyes, and skull) grows very rapidly from birth to 6 years, when it reaches 90 per cent of adult size. Reproductive organs remain small until puberty when they grow rapidly to adult size.

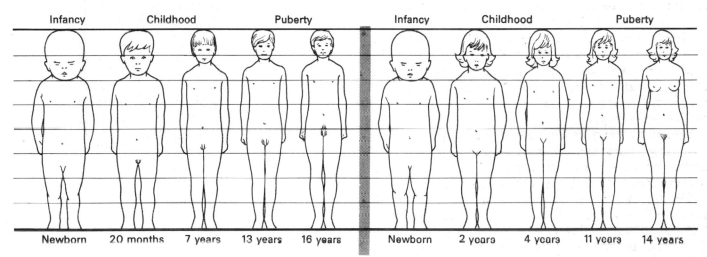

Fig. 3. **Changes in body proportions.** Compare body proportions of babies and adults. Note that proportions change at different rates in boys and girls.

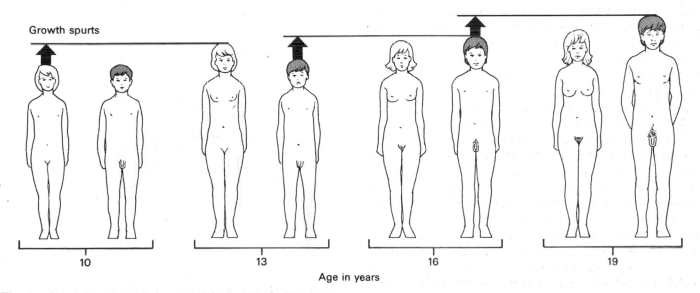

Fig. 4. **Growth spurts** in boys and girls. Rapid growth is followed by a period of 'filling-out'.

65

Birth control

Birth control allows people to limit the size of their families by either preventing fertilization or implantation, or by limiting sexual intercourse to times when fertilization is unlikely to occur.

Preventing conception

Withdrawal The man withdraws his penis just before ejaculation occurs. The idea is to avoid shedding sperms into the woman but this method is extremely unreliable because many sperms are shed without the man's knowledge some time before ejaculation.

The condom A condom is a sheath of thin strong latex rubber which is unrolled onto the erect penis before intercourse. Sperms are trapped inside the condom and so do not enter the woman. This method is quite reliable, especially if used in conjunction with spermicides.

Spermicides These consist of a jelly, cream, or foam containing chemicals which kill sperms. Spermicides are inserted into the vagina before intercourse but, since they are unreliable on their own, they should be used in conjunction with a condom or diaphragm.

Diaphragm This is a dome-shaped disc of thin rubber which is kept in shape by a spring around the rim. A diaphragm is very reliable provided the woman obtains one which exactly fits her cervix. It must be smeared with a spermicide before fitting and it should be fitted half an hour before intercourse. The diaphragm should not be removed until six to eight hours after intercourse.

Intra-uterine device (IUD) An IUD is a piece of flexible plastic in the form of a loop or coil which is inserted into the womb. It must be fitted by a doctor, and regular medical checks are necessary to ensure that the device remains in place and does not irritate the lining of the womb. This method is very reliable. It appears to work by preventing a fertilized ovum from becoming implanted in the womb.

Sterilization This involves a surgical operation. In men the operation is called **vasectomy**, and involves cutting, tying, or blocking the tubes which carry sperms from the testes to the penis. In women the operation involves cutting, tying, or blocking the fallopian tubes. The operation is effective immediately in women, but men can still produce live sperms for up to six months afterwards.

Contraceptive pills These contain one or more female hormones. They prevent ovulation, stop the mechanism which transports ova along the fallopian tubes, and cause mucus in the entrance to the womb to become sticky so that sperms are unlikely to swim through. Pills are almost 100 per cent reliable when taken according to instructions. But they must not be taken by women with liver diseases or diabetes, and in rare instances they have been known to cause blood clots. There is evidence that women who smoke and take contraceptive pills are more likely than non-smokers to develop blood clots, migraine headaches, and allergies to certain foods.

Avoiding conception

The rhythm, or mucothermic, method This method involves avoiding intercourse during the time each month when the woman is fertile. This time can be discovered in two ways. As soon as her menstrual bleeding stops the woman must make daily observations to discover when a thin clear mucus is discharged from the entrance to her vagina. In addition she must record her temperature very carefully every day and note when it goes up by $0.1\,°C$ to $0.5\,°C$ and stays up for several days. Ovulation occurs soon after the mucus appears and it is followed, within two or three days, by a slight rise in temperature. Consequently intercourse should be avoided after the mucus appears and should not start again until at least three days after the temperature rise occurs.

The temperature rise at ovulation is very small indeed. Consequently this method is only reliable when an accurate clinical thermometer is used correctly, and very careful records are kept.

A **A condom** is a sheath which is unrolled over the penis before intercourse.

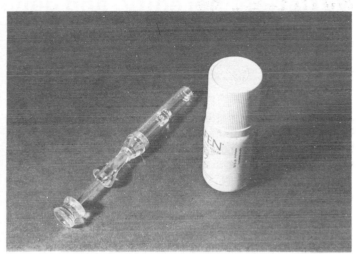

B **Spermicides** are chemicals which kill sperms. They are produced in tablet form (pessaries) or as a jelly. Both types are inserted into the vagina before intercourse (a plunger is used to insert the jelly).

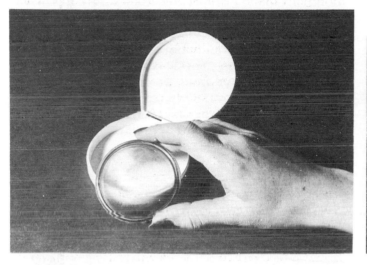

C **A diaphragm** is a dome-shaped disc which is fitted over the cervix before intercourse.

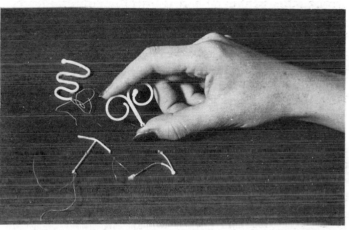

D **Intra-uterine devices** are inserted into the womb.

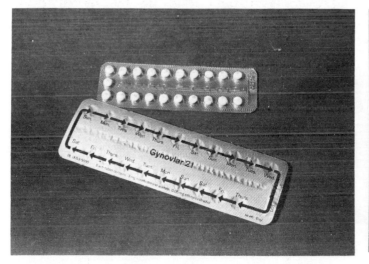

E **Contraceptive pills** contain one or more female hormones which, among other things, prevent ovulation. A pill is taken every day for 21 days, then they are discontinued for the next seven days to allow a 'period' (menstruation) to occur.

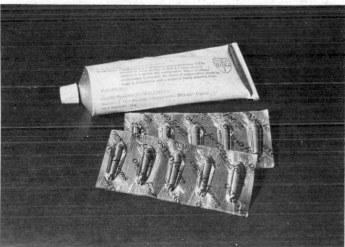

F **Pessaries.** One of these is inserted into the vagina before intercourse. The pessary dissolves, releasing spermicide chemicals.

66

Variation, heredity, and genetics

People throughout the world have the same general body shape, and share the same set of organs, but no two people are exactly alike. Even 'identical' twins differ in some ways.

Features such as height, weight, hair and eye colour, shape of the face, language, habits, intelligence, and skills differ from person to person. Differences between these and other characteristics are examples of **variation**.

Types of variation

People occur in so many different sizes that it is possible to arrange even a small group into a continuous line from the shortest to the tallest (Fig. 1), or from the lightest to the heaviest. Characteristics like these, in which there are many intermediate forms between two extremes, are said to show **continuous variation**. Intelligence also shows continuous variation (Fig. 2).

Other characteristics have no, or very few, intermediate forms. For example, with rare exceptions people are either male or female and not inbetween the two, and they can either roll their tongue into a U-shape or they cannot (Fig. 3A). These are examples of features showing **discontinuous variation**.

Inherited and acquired characteristics

People inherit some characteristics from their parents. For example, a child may inherit its hair and eye colour from one or other of its parents, together with features such as the shape of its nose, ears, mouth, its blood group, and certain disorders such as colour blindness. These are examples of **hereditary characteristics**. They are fixed from the moment when a sperm and an ovum fuse together to form a fertilized egg. The scientific study of how hereditary characteristics pass from parents to the young is called **genetics**.

Features such as language, scars, skills, and habits are called **acquired characteristics** because they are acquired after birth.

Fig. 1. **Continuous variation** is shown by characteristics such as height. There are so many intermediate sizes between two extremes that even a small group of people can be arranged in a continuous line from the shortest to the tallest.

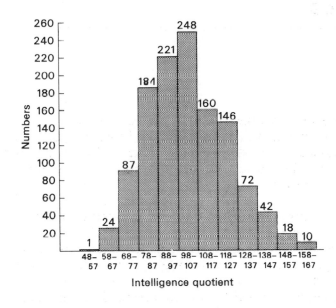

Fig. 2. **Intelligence** also shows continuous variation. Note that the majority of children given this intelligence test scored around the average mark (100), while only a few scored very low or very high marks. A distribution of this type is typical of characteristics showing continuous variation.

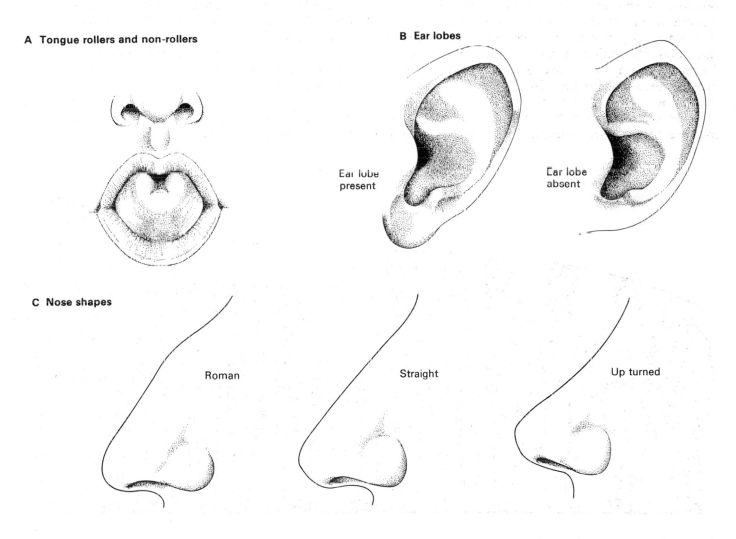

A Tongue rollers and non-rollers

B Ear lobes

Ear lobe present

Ear lobe absent

C Nose shapes

Roman

Straight

Up turned

Fig. 3. **Discontinuous variation** is shown by characteristics with no, or very few, intermediate forms. For example, people can either roll their tongues or they cannot (**A**), and noses and ears usually occur in one of a few distinct shapes (**B** and **C**).

67

Chromosomes, genes, and mitosis

Genes can be described as 'units of inheritance' because each gene controls the development of one or a set of hereditary characteristics. Genes are not confined to controlling characteristics like eye colour or the shape of a nose. Each cell of a human body contains tens of thousands of genes and they are responsible for the development of all the tissues and organs as well as external features.

What are genes made of and where are they situated in an organism? Genes are made of complex chemicals described in Units 71 and 72, and they form part of objects called chromosomes, which make up the bulk of a cell nucleus.

Humans and nearly all other multicellular organisms begin life as a single cell: a fertilized egg cell or zygote. A human zygote contains 46 chromosomes and between them they contain all the instructions for building a complete human being.

A human body is built from a zygote by cell division of a type called **mitosis**. During mitosis the zygote divides in such a way that the two daughter cells produced from it have the same number of chromosomes as the zygote itself (Fig. 1 describes how this happens). Unit 71 describes how, just before a cell divides by mitosis, its chromosomes duplicate themselves, producing two identical sets of chromosomes. At mitosis one set of chromosomes passes into each daughter cell. Then these two cells divide by mitosis producing more cells which do the same so that, as the body grows, every cell receives 46 chromosomes which are an exact copy of the chromosomes in the original zygote, complete with a set of building instructions in the form of genes.

As growth proceeds instructions in the genes ensure that cells become tissues, that tissues become organs and, eventually, that individual features such as eye and hair colour appear.

Mitosis does not stop when the body reaches adult size. Mitosis in the epidermis produces cells which replace those worn away from the surface of the skin; it is responsible for producing new red and white blood cells in the bone marrow; it produces growth of hair and finger-nails; and it creates new cells to repair damaged tissues.

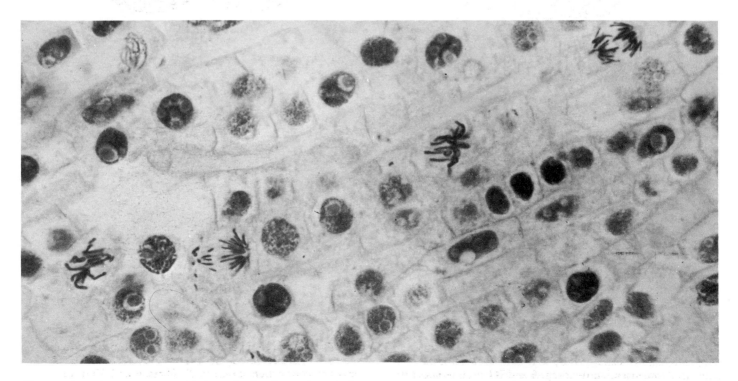

Chromosomes in the dividing cells from the tip of an onion root. Using the Figure opposite as a guide, identify the stages of mitosis.

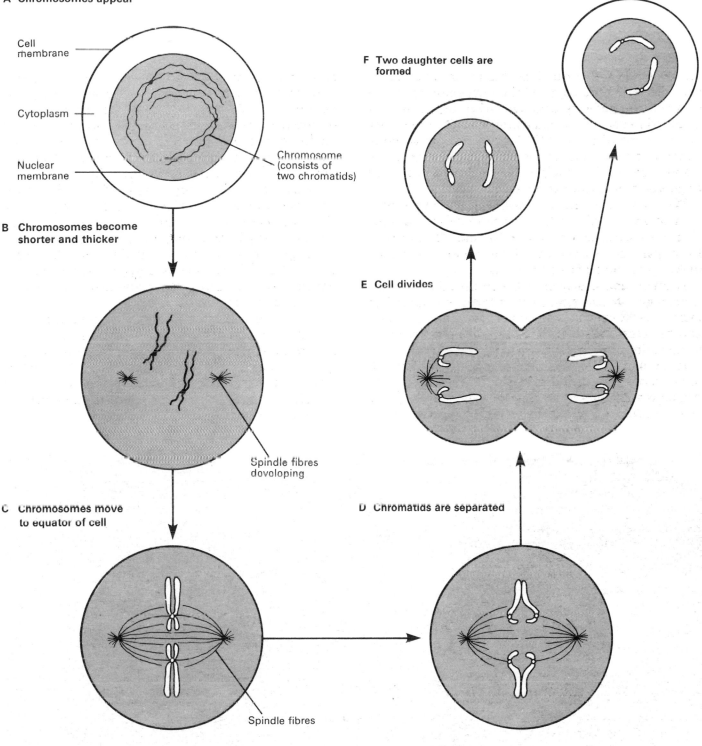

A **Chromosomes appear**

Cell membrane

Cytoplasm

Nuclear membrane

Chromosome (consists of two chromatids)

B **Chromosomes become shorter and thicker**

Spindle fibres developing

C **Chromosomes move to equator of cell**

Spindle fibres

F **Two daughter cells are formed**

E **Cell divides**

D **Chromatids are separated**

Fig. 1. Cell division by mitosis produces cells with the same number of chromosomes as the parent cell. Chromosomes are only visible when a cell is dividing. **A** and **B** Chromosomes first appear as long, fine double threads which become shorter and thicker. **C** The chromosomes move to the middle (equator) of the cell where they become attached to fine fibres called the spindle. **D** Each chromosome separates into two parts which move to opposite ends of the cell. They are probably pulled apart by contraction of the spindle fibres. **E** and **F** The cell divides, separating the two groups of chromosomes which then form a nucleus in each daughter cell.

135

68

Gametes, meiosis, and sex chromosomes

Formation of gametes by meiosis

Humans have 46 chromosomes in every body cell. But if **gametes** (sperms and ova) had this number, fertilization would produce young with 92 chromosomes in their cells, and the number would double every time reproduction occurred.

This does not happen because gametes are produced by a type of cell division called **meiosis**. Unlike mitosis, described in Unit 67, meiosis produces four daughter cells with *half* the normal number of chromosomes. For this reason meiosis is called **reduction division**.

Human sperms and ova have 23 chromosomes. This is called the **haploid number** of chromosomes, and the normal or **diploid number** of 46 chromosomes is restored at fertilization. These 46 chromosomes are, in fact, 23 pairs. In technical terms they are called **homologous pairs**, partly because one member of each pair is the same shape and size as the other. Sex chromosomes, described below, can be an exception to this rule. The description of meiosis in Figure 2 should be studied before reading any further.

Genes and homologous chromosomes

The movement of chromosomes during meiosis and fertilization is clear evidence that they contain the units of heredity, called **genes**.

1. Unit 70 explains that genes occur in pairs. During meiosis chromosomes form homologous pairs.

2. Gene pairs separate at gamete formation (meiosis) and one member of each pair enters each gamete. Homologous chromosomes behave in exactly the same way during meiosis.

3. Gene pairs and homologous pairs of chromosomes are restored at fertilization.

Sex chromosomes

The 23 pairs of chromosomes in human cells include a pair of sex chromosomes, so-called because they determine whether a zygote develops into a male or a female. There are two types of sex chromosome: the X and the Y types. Y chromosomes are smaller than X chromosomes.

A zygote containing a pair of X chromosomes develops into a female, but a zygote with an X and Y chromosome develops into a male. Figure 1 explains how the two types of zygote are produced.

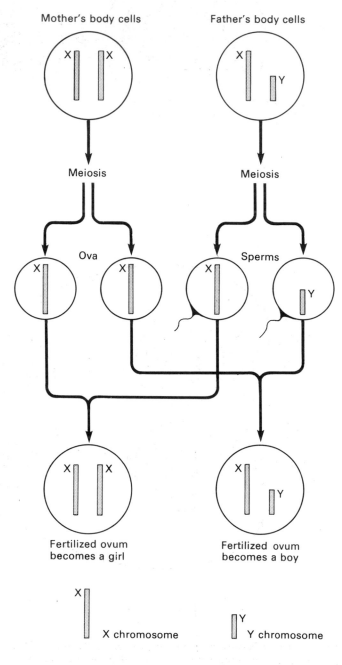

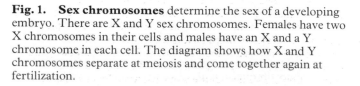

Fig. 1. **Sex chromosomes** determine the sex of a developing embryo. There are X and Y sex chromosomes. Females have two X chromosomes in their cells and males have an X and a Y chromosome in each cell. The diagram shows how X and Y chromosomes separate at meiosis and come together again at fertilization.

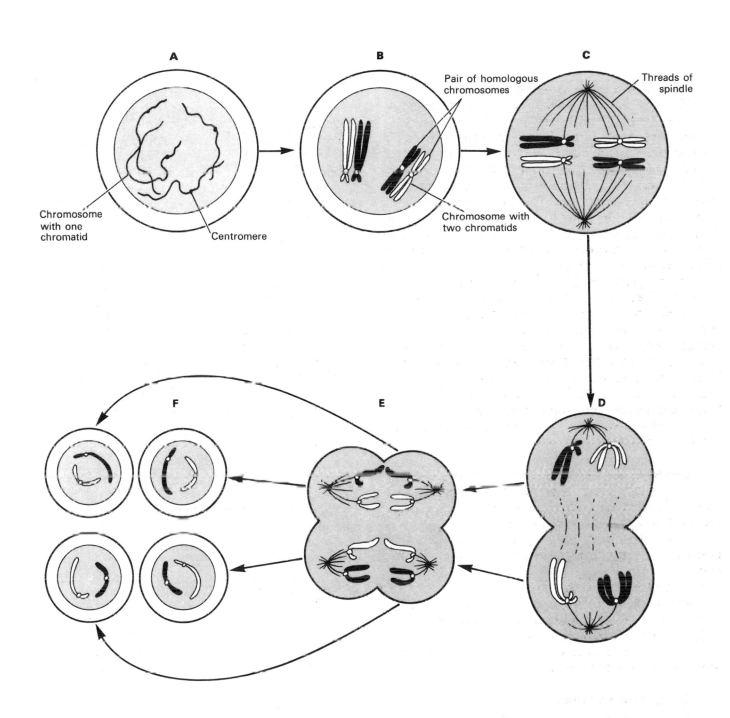

Fig. 2. Cell division by meiosis produces cells with half the number of chromosomes as the parent cell. Meiosis occurs in the testes and ovaries producing sperms and ova with 23 chromosomes each. **A** Chromosomes become shorter and thicker. **B** Chromosomes form homologous pairs. (One of each pair is drawn in black to distinguish it from its partner.) Each chromosome forms a second strand. **C** The homologous pairs become arranged along the equator of the cell, attached to spindle fibres. **D** The members of each homologous pair of chromosomes separate and move in opposite directions. The cell begins to divide into two. **E** Each chromosome splits into two parts which move in opposite directions. **F** The cell divides into four parts each containing half the number of chromosomes possessed by the original cell. In the testis the four new cells develop into sperms. In the ovary only one of the four develops into an ovum. The remaining three are very small and soon disintegrate.

69

Heredity and genes

For many reasons it is difficult, and often impossible, to use humans in the experimental study of heredity. It is far easier to experiment with fast-breeding organisms such as the tiny fruit fly *Drosophila melanogaster*.

To avoid confusing results it is necessary to study the inheritance of only one or two characteristics at a time. *Drosophila* is ideal for this purpose since it occurs in varieties with clearly visible differences, such as red and white eyes, and short (vestigial) and normal wings (Fig. 1).

Characteristics like these show discontinuous variation, explained in Unit 66. Therefore, in studying the inheritance of wing length, for example, normal-winged flies can be mated, or **crossed**, with a short-winged variety and the young will have either normal or short wings with no intermediate forms.

It is vital that each variety of organism used in heredity experiments is a **pure line**. Pure line normal-winged *Drosophila* flies for example will, when bred together, produce only normal-winged young. Their 'line' is 'pure' in the genetic sense.

When organisms which differ in some way are bred together their young are called **hybrids**. When the parent organisms differ in only one way (such as wing length) the experiment is called a **monohybrid cross**. Figure 1 shows the result of a monohybrid cross between pure line normal-winged *Drosophila* flies and pure line short-winged flies. Their young, called the **first filial generation**, F_1, are collected and their wing length observed.

Results of a monohybrid cross

Without exception all the F_1 flies of this cross will have normal wings. This will happen because the ability to produce normal wings is an example of a **dominant characteristic**; that is, it 'dominates' the short wing characteristic.

If the F_1 flies are bred together to produce an F_2 generation, young of two types will be produced. Roughly three-quarters of the F_2 flies will have normal wings and one-quarter will have short wings, giving a ratio of 3:1. Since the short wing characteristic has reappeared after 'receding' in the F_1 generation it is called a **recessive characteristic**.

Explaining these results

It is now known that hereditary characteristics, not only in *Drosophila* but in all organisms, are controlled by microscopic structures called genes. Unit 67 explains that genes form part of objects called chromosomes which are situated in the nucleus of every cell in an organism's body.

The results of a monohybrid cross can be explained by using letters of the alphabet to represent the genes involved (Fig. 2). A dominant gene, like the one which produces normal wings in *Drosophila*, can be represented by a capital **A** and a recessive gene, like the one which produces short wings in *Drosophila*, can be represented by a small **a**.

Genes work together in pairs. A pure line normal-winged *Drosophila* has the genes **AA** in every cell, whereas the cells of a short-winged fly have the genes **aa**. Meiosis in a fly's reproductive organs produces sperms and ova with only *one* gene from each pair. So sperms of an **AA** male get only one **A** gene each, while the ova of an **aa** female get one **a** gene each.

At fertilization the genes come together and form pairs again. **A** sperms fertilize **a** ova making F_1 hybrids with **Aa** genes in their cells. The **A** gene will dominate the **a** gene, which explains why all F_1 flies have normal wings.

F_1 males produce equal numbers of **A** and **a** sperms, and F_1 females produce equal numbers of **A** and **a** ova. The matrix at the bottom of Figure 2 shows how random fertilization amongst these sperms and ova produces an F_2 generation in which the dominant and recessive characteristics appear in the ratio of 3:1.

Summary

Hereditary characteristics are controlled by genes, and a monohybrid cross reveals that there are dominant genes and recessive genes.

Genes occur in pairs, but when sex cells are produced the members of each pair separate so that sperms and ova receive only one of each. Fertilization brings the gene pairs together again, often in a different order. The cells of a pure line organism contain either two dominant or two recessive genes for a particular characteristic. Hybrid organisms have one dominant and one recessive characteristic.

Experiments like the one described here are important because they reveal the natural laws of inheritance. This knowledge helps plant and animal breeders to improve crops and farm livestock, and helps doctors to understand hereditary disorders.

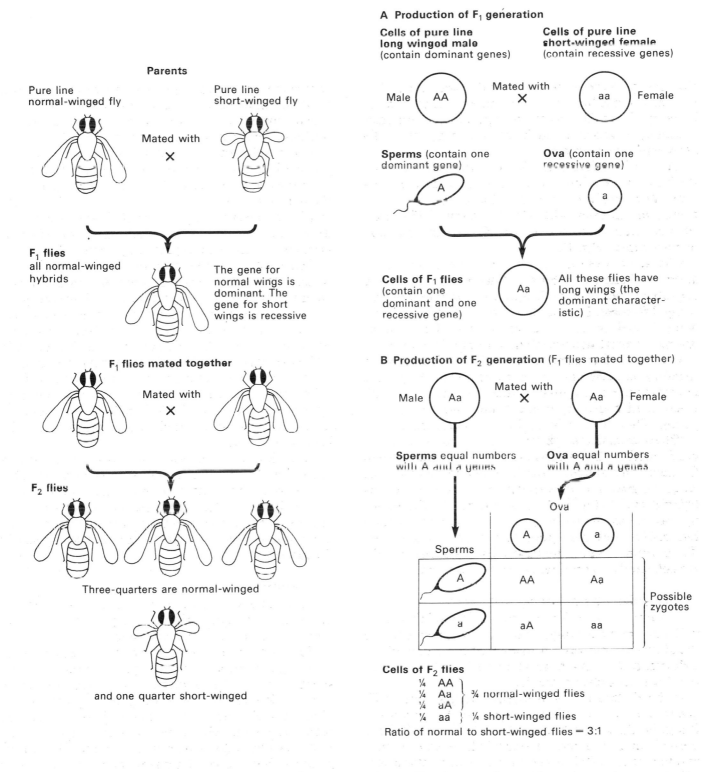

A Production of F₁ generation

Cells of pure line long winged male (contain dominant genes)   **Cells of pure line short-winged female** (contain recessive genes)

Male AA   Mated with ✕   aa Female

Sperms (contain one dominant gene)   **Ova** (contain one recessive gene)

A   a

Cells of F₁ flies (contain one dominant and one recessive gene)   Aa   All these flies have long wings (the dominant characteristic)

B Production of F₂ generation (F₁ flies mated together)

Male Aa   Mated with ✕   Aa Female

Sperms equal numbers with A and a genes   **Ova** equal numbers with A and a genes

Ova

Sperms

	A	a	
A	AA	Aa	Possible zygotes
a	aA	aa	

Cells of F₂ flies

¼ AA
¼ Aa } ¾ normal-winged flies
¼ aA
¼ aa } ¼ short-winged flies

Ratio of normal to short-winged flies = 3:1

Parents

Pure line normal-winged fly   Pure line short-winged fly

Mated with ✕

F₁ flies all normal-winged hybrids   The gene for normal wings is dominant. The gene for short wings is recessive

F₁ flies mated together

Mated with ✕

F₂ flies

Three-quarters are normal-winged

and one quarter short-winged

Fig. 1. A monohybrid cross between normal-winged and short-winged fruit flies *Drosophila melanogaster* (differences between males and females not shown). The F₁ generation all have normal wings, which shows that this is a dominant characteristic and that short wings is a recessive characteristic. In the F₂ generation normal and short wings occur in the ratio of 3:1. This is explained in Fig. 2.

Fig. 2. Dominant and recessive genes can be represented by the letters **A** and **a** respectively. Genes occur in pairs in body cells but sperms and ova contain only one member of each pair. Gene pairs come together again at fertilization but often in a different order.

70

Some technical terms used in genetics

Homozygous and heterozygous gene pairs

Genes operate in pairs, and each pair controls one or a set of hereditary characteristics. When the members of a particular gene pair are identical the pair is said to be **homozygous**. A gene pair can be homozygous dominant (**AA**) or homozygous recessive (**aa**).

The members of a gene pair can be different: one dominant and the other recessive (**Aa**). In this case the pair is said to be **heterozygous**.

The locus and alleles of a gene

The position of a gene on a chromosome is called the **locus** of that gene. Unit 68 explains that chromosomes occur in pairs, called homologous chromosomes. Each pair of homologous chromosomes holds a particular set of genes, arranged so that the members of each gene pair occupy the same locus on both homologous chromosomes.

Genes which occupy the same locus on homologous chromosomes are said to be **alleles** of each other. Inheritance of blood groups, for example, is controlled by three alleles: the genes A, B, and O. But, since genes operate in pairs, only two of these alleles can be present at the same time in an individual's cells. Gene O is recessive to both A and B, but A and B are said to be **co-dominant** because neither can 'dominate' the other. Table 5 shows how the different combinations of these alleles produce the various blood groups.

Genotype and phenotype

The set of genes possessed by an organism is referred to as the **genotype** of that organism. The word genotype can also refer to one particular set of genes, such as those controlling eye colour. In contrast, the word **phenotype** refers to the actual visible physical characteristics of an organism's body.

For example, two organisms can have different genotypes but the same phenotype. Wing length in *Drosophila* is controlled by genes which can be represented by **A** and **a**. Genotypes **AA** and **Aa** are different but produce flies with the same phenotype: normal (long) wings. This is because **A** is dominant. On the other hand two organisms with the same genotype can have different phenotypes. 'Identical' twins, for instance, have exactly the same genotype because they developed from the same zygote. As the twins grow, however, interaction between these genes and their environment can result in different phenotypes. In particular they may differ in height, weight, and intelligence, especially if they are separated at an early age and reared in different environments.

Mutation

A mutation is a sudden change in a gene or chromosome which alters the way in which it controls development. Mutations are usually harmful, but are quite rare. Mutations occur more frequently in organisms exposed to X-rays, gamma rays, and nuclear radiation. The most important mutations are those in cells which produce gametes (sex cells) because these mutations can pass from parents to young.

Albinos, for example, result from a mutation in the genes which control skin colour (see photograph opposite). **Haemophilia** is a disease which results from mutation in a gene which controls the formation of the blood-clotting mechanism. The blood of a haemophiliac fails to clot in wounds. Inheritance of haemophilia is sex-linked.

Sex-linked inheritance

Certain types of haemophilia and colour blindness are examples of sex-linked characteristics, so-called because they are more likely to occur in males than females. There are several reasons why this happens. The genes for haemophilia and colour blindness occur on X chromosomes, and both genes are recessive to a corresponding dominant allele for normal blood clotting and colour vision. Females have two X chromosomes. Consequently a gene for colour blindness or haemophilia on one X chromosome will, in most cases, be masked by a dominant allele on the other X chromosome. Very few females have the recessive genes on both X chromosomes.

Males have one X and one Y chromosome in their cells. The Y chromosome is small and lacks both blood-clotting and colour-vision genes. If, therefore, a male has received from his mother an X chromosome with a gene for colour blindness or haemophilia he has no corresponding dominant allele and will suffer from the disorder.

Figure 1 illustrates the passage of sex-linked genes from parents to young.

Table 5. Inheritance of blood groups.

Genes	Blood groups
OO	O
AA	A
AO	A
BB	B
BO	B
AB	AB

O is recessive to A and B but A and B are co-dominant—neither is dominant to the other.

An albino and a normal dark-skinned African. Albinism results from a mutation in the genes which control skin coloration. The skin of an albino is pink and almost transparent and is quickly damaged by exposure to sunlight. The mutation is recessive.

Key   X X chromosome with normal blood-clotting gene
       X̄ X chromosome with haemophilia
       Y Y chromosome

A   Haemophiliac man   Normal woman

          X̄Y   XX
        Meiosis   Meiosis

                        Ova

Possible zygotes:

Sperms	X	X	
X̄	X̄X	X̄X	½ X̄X Girl 'carriers' of haemophilia
Y	XY	XY	½ XY Normal boys

B   Normal man   Carrier woman

          XY   X̄X
        Meiosis   Meiosis

                        Ova

Possible zygotes:

Sperms	X̄	X	
X	X̄X	XX	¼ X̄X Girl carriers / ¼ XX Normal girls
Y	X̄Y	XY	¼ X̄Y Haemophiliac boys / ¼ XY Normal boys

Fig. 1.   Inheritance of haemophilia is sex-linked.
Haemophilia genes pass from a father to his daughters, but from a mother to her sons and daughters. Sons who receive the gene are haemophiliacs, daughters are carriers unless, very rarely, they receive two recessive genes.

71

DNA and the genetic code

Genes are the instructions which direct the growth and development of organisms. The popular name for these instructions is the **genetic code**. What is the genetic code made of and how are its instructions obeyed? This Unit and the next try to answer these questions.

In 1944 it was shown that genes are made of a chemical called **deoxyribonucleic acid**, or **DNA**. In the early 1950s the structure of DNA was discovered and this led to an understanding of the genetic code.

Structure of DNA

DNA is a long, thread-like molecule similar in shape to a rope ladder twisted into a spiral (Fig. 1). The upright sides of the ladder are made of alternate sugar and phosphate molecules joined into a chain. The rungs of the ladder are made of chemical bases, of which there are four types: adenine, guanine, cytosine, and thymine. For simplicity these bases are usually referred to by their initial letters A, G, C, and T (Fig. 2).

The genetic code

The genetic code is formed by the sequence in which these four chemical bases are arranged along the length of a DNA molecule. Put another way, if the genetic code were a set of written rather than chemical instructions, the letters A, G, C, and T would be the alphabet in which the 'words' of the code were written.

In fact, the code is written in words of only three letters each; that is, it consists of groups of three adjacent bases. For example, the base sequence AAT along one strand is one word, CAA is another, and so forth. The number of different three-letter words which can be composed out of the four letters A, G, C, and T is 64.

It is now known that these three-letter words (groups of three bases) are the instructions to a cell telling it how to make protein molecules out of amino acids. The sequence AAT (adenine, adenine, thymine) in the code tells a cell how to make the amino acid leucine, and CAA (cytosine, adenine, adenine) tells a cell how to make the amino acid valine.

A single gene consists of a thousand or more groups of three bases. Together these groups represent instructions to make one complete protein molecule of a specific type. A cell 'obeys' these instructions by putting together the correct amino acids in the exact sequence directed by the gene. The next Unit explains how this happens.

DNA, therefore, controls the types of protein manufactured in cells. In this way DNA controls the structure of an organism because proteins are the building materials out of which cells, tissues, and organs are made. But equally important is the fact that DNA directs the chemical activities which take place in an organism, because some proteins are enzymes and enzymes control the speed and type of chemical reactions in cells.

Duplication of DNA during cell division

Unit 67 explains how cell division by mitosis produces daughter cells with exact copies of the chromosomes present in the original cell. Before mitosis can occur in a cell the DNA molecules in its chromosomes must be duplicated exactly. Eventually these duplicates form a second set of chromosomes.

Before describing DNA duplication it is necessary to look more closely at its structure. The rungs of the ladder-like DNA molecule are made up of chemical bases represented by A, G, C, and T. Each rung consists of two bases joined at the centre of the ladder (Fig. 2). There are millions of rungs in one DNA molecule but in every one, base A is *always* opposite base T, and base C is *always* opposite base G. This is known as the **pairing rule**.

When DNA duplicates, the base pairs in each rung separate and the molecule comes apart into two strands as if it had been unzipped up the middle (lower half of Fig. 2). The two single strands are now built up into double strands again using new bases supplied by the cell. The pairing rule ensures that the base T is always joined up with an A (or an A with a T) and the base C is always joined up with a G (or a G with a C) as shown in Figure 2. The result is two DNA molecules which are identical to the original.

Very rarely an error occurs during duplication of DNA and results in a slight change in the genetic code. This is one way in which mutations occur.

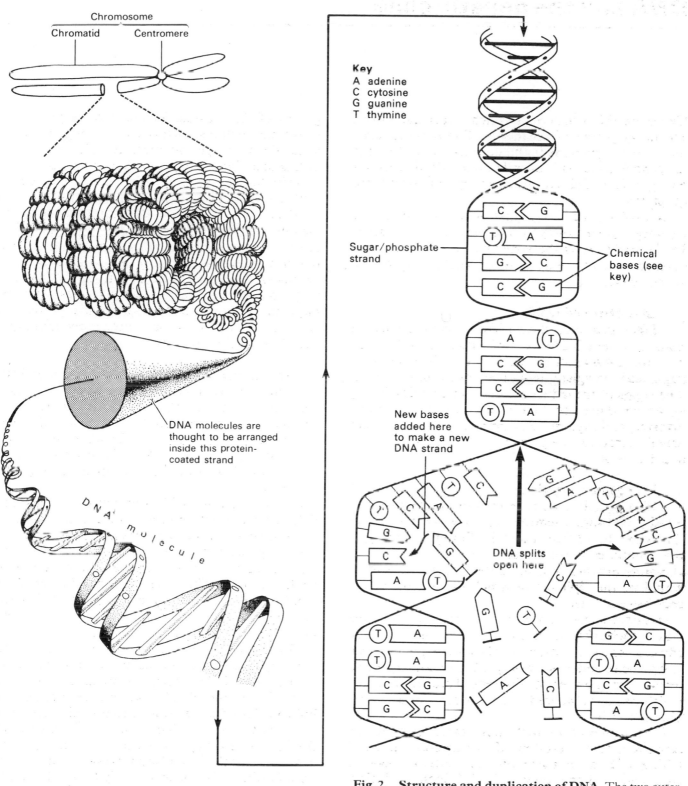

Fig. 1. **Chromosomes and DNA.** A chromosome consists of strands of protein which coil up as shown just before cell division. It is possible that DNA molecules run through the centre of these protein strands. A DNA molecule consists of two outer filaments with cross-pieces between them — like a ladder twisted into a spiral.

Fig. 2. **Structure and duplication of DNA.** The two outer strands of a DNA molecule are made of alternate sugar and phosphate molecules. The cross-pieces consist of chemical bases: **A** (adenine), **G** (guanine), **C** (cytosine), and **T** (thymine). **A** is always opposite **T**, and **C** is always opposite **G** Duplication of DNA occurs just before cell division. The molecule splits into two strands and new chemical bases are added to these, building up a new double strand again. The result is two DNA molecules identical to the original one.

72

Transcribing the genetic code

The previous Unit explains that each gene on a DNA molecule carries instructions for making a particular type of protein. This Unit describes how these instructions are obeyed or, in technical terms, how the genetic code is **transcribed**. This Unit also explains how each cell transcribes only those genes concerned with its particular functions leaving the thousands of other genes in its chromosomes inactive.

Most of a cell's DNA is contained in the chromosomes in its nucleus, but proteins are made in the cytoplasm of a cell. Somehow the coded instructions for protein manufacture must pass from the nucleus to the cytoplasm and then be transcribed. This task is carried out by two types of a substance called **ribonucleic acid**, or **RNA**, which work in conjunction with microscopic granules in the cytoplasm called **ribosomes**. The two types are called messenger RNA and transfer RNA. Figure 1 and the following notes describe the commonly accepted theory of how this transcription happens.

A DNA molecule opens up

The region of DNA containing the gene for the protein required by the cell opens out, exposing the sequence of chemical bases with the necessary coded instructions (Fig. 1A).

Messenger RNA copies the code

The exposed single strand of DNA with the coded instructions is built up into a double strand again in a similar fashion to DNA duplication described in Unit 71, except that this time the new second strand is made of messenger RNA.

The main difference between RNA and DNA is that in RNA the base thymine is replaced by the base uracil, which is represented by the letter U. Consequently the messenger RNA strand is built up according to the pairing rule: A (adenine) opposite U (uracil), and G (guanine) opposite C (cytosine). In this way the messenger RNA molecule becomes an exact copy of the exposed gene's coded instructions on the DNA molecule.

Next, messenger RNA leaves the nucleus and moves into the cytoplasm where it becomes attached to a ribosome (Fig. 1B). This is where protein building takes place.

The cytoplasm of a cell contains millions of amino acids—the building blocks from which proteins are made. The human body requires 20 different amino acids to build its proteins, and each type of protein molecule contains a different number of amino acids arranged in a different sequence.

Transfer RNA collects the amino acids

It is the function of transfer RNA molecules to move around the cytoplasm collecting the correct types of amino acid (Fig. 1C). Transfer RNA then transfers the amino acids to messenger RNA on a ribosome, where the amino acids are linked up in the correct sequence to make a protein molecule.

A transfer RNA molecule contains three chemical bases, and Unit 71 explains that a group of three such bases is the code for a particular amino acid. Each transfer RNA molecule picks up a single amino acid molecule of the type directed by its part of the code, and floats around the cytoplasm until it reaches a ribosome which has a strand of messenger RNA attached to it.

Amino acids are linked together

When a transfer RNA molecule (complete with its amino acid) arrives at a ribosome its three chemical bases join up with a corresponding set of three bases on the messenger RNA molecule, again using the pairing rule A opposite U and G opposite C. As this is happening the amino acid carried by the transfer RNA links up with amino acids on an adjacent transfer RNA molecule. Repetition of this process builds up a chain of amino acids until a complete protein molecule is formed (Fig. 1D).

Summary

Ribosomes, together with messenger and transfer RNA, form an assembly line for proteins, and DNA is the library where assembly instructions are stored.

Messenger RNA molecules visit the DNA library, copy the instructions for making a protein, and carry them to a ribosome in the cytoplasm. Transfer RNA molecules collect amino acids and transfer them to the ribosome. Here, the amino acids are fitted together according to the instructions built into the structure of the messenger RNA.

The sequence in which the amino acids are fitted together depends upon the sequence of bases in the messenger RNA, which depends upon the sequence of bases in the region of DNA which was copied.

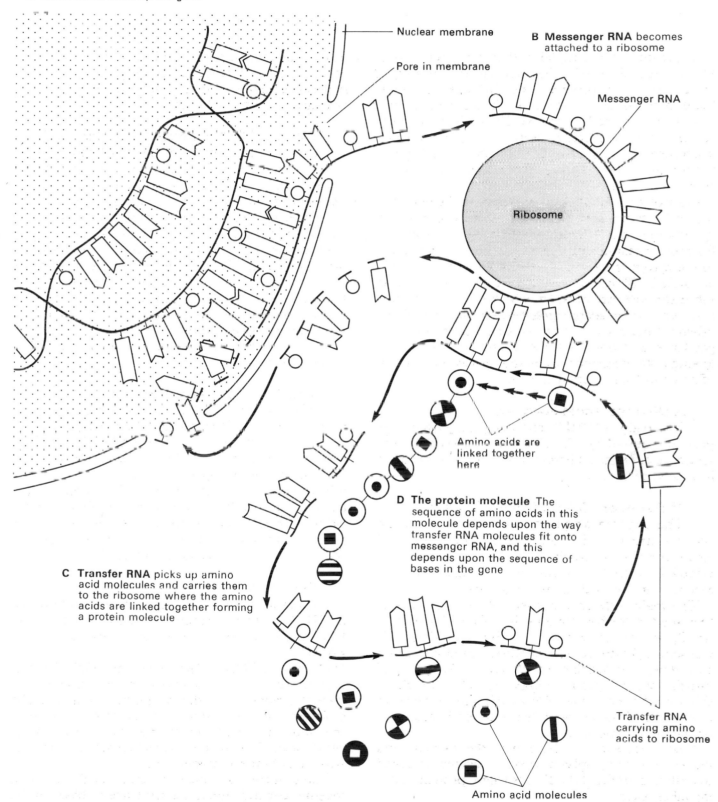

A In the nucleus part of a DNA molecule opens up exposing the gene for a particular protein. Messenger RNA copies the sequence of bases which make up the gene

Nuclear membrane

Pore in membrane

B Messenger RNA becomes attached to a ribosome

Messenger RNA

Ribosome

Amino acids are linked together here

D The protein molecule The sequence of amino acids in this molecule depends upon the way transfer RNA molecules fit onto messenger RNA, and this depends upon the sequence of bases in the gene

C Transfer RNA picks up amino acid molecules and carries them to the ribosome where the amino acids are linked together forming a protein molecule

Transfer RNA carrying amino acids to ribosome

Amino acid molecules

Fig. 1.   Transcription of the genetic code.

73

Some causes of disease

A healthy person is one whose mental, physical, and chemical processes are working efficiently and harmoniously. Many things can upset this harmony and cause 'dis-ease'.

Organisms which cause disease

Diseases are often caused by organisms which live as **parasites**. A parasite is an organism which obtains its food from the living body of another organism called the **host**; but not all parasites harm their host. Most humans, for example, have a parasitic bacterium, *Escherischia coli*, living harmlessly in their intestines. Parasites which harm their hosts are said to be **pathogenic**, or **pathogens**.

Pathogenic micro-organisms

The words **micro-organism** and **microbe** refer to microscopic creatures including viruses, bacteria, rickettsias, protozoa, and fungi such as yeasts and moulds. The majority of microbes are harmless, and some are very useful, such as the yeasts used in baking and wine-making and the bacteria and fungi used to make antibiotics.

Pathogenic microbes are commonly known as germs. Examples are: viruses which cause influenza; bacteria which cause food poisoning and cholera; protozoa which cause dysentery and malaria; and fungi which cause ringworm and athlete's foot. Some of these microbes are described in Unit 74.

Parasitic worms and insects

Roundworms (nematodes) and flatworms (platyhelminths) which live as parasites are described in Units 75 and 76. Insects such as fleas and lice (described in Unit 77) can live as parasites but although they cause discomfort they are far more important as the carriers, or **vectors**, of pathogenic microbes. Female anopheline mosquitoes, for instance, are vectors of the microbes which cause malaria.

How pathogens cause disease

Figure 1 describes some of the ways in which pathogenic organisms enter the body. Once inside, some pathogens release poisonous chemicals called **toxins**. Most toxins are proteins and are by-products of the parasites' metabolism. They produce disease symptoms in the host like high temperature, headache, and vomiting. Symptoms do not appear immediately a pathogen enters a host. There is an interval called the **incubation period** before symptoms appear, during which germs multiply rapidly or larger parasites develop to full size.

Some parasites, including certain nematode worms and insects, bore through host tissues causing wounds which may then become infected by bacteria and viruses. This is called **secondary infection**.

Epidemic and endemic diseases

When a disease is always present in an area, such as malaria in parts of Africa, it is said to be **endemic** to that region. A sudden outbreak of the disease which spreads through a population is called an **epidemic**, and if it spreads through several countries the disease has become **pandemic**.

Notifiable diseases

Doctors are compelled to report an outbreak of certain diseases to their Local Health Authority. These are called **notifiable diseases** and include: smallpox, scarlet fever, diphtheria, and typhoid fever. After notification, arrangements are made to prevent the spread of infection. Patients are isolated and their relatives, friends, and others who have been in contact with them are vaccinated.

People who have been in contact with an infectious disease are kept in **quarantine** for a time. This means they are isolated in a hospital for a few days longer than the incubation period of the disease to reduce risk of spreading the infection.

Other causes of disease

Parasites are not the only cause of disease. Many diseases such as haemophilia (described in Unit 36) are inherited. Diabetes results from defects in certain chemical processes in the body; and some mental illnesses, like anxiety states, can develop in people under great emotional stress.

Disease is also caused by a poor diet. Vitamin and mineral deficiency diseases are discussed in Unit 21. **Kwashiorkor** is a disease caused by lack of protein, which results in cracked skin and damage to the liver. Lack of fibre (bran) in the diet does not cause a deficiency disease but can contribute to the development of bowel cancer, piles, appendicitis, and constipation. There is also evidence that excess animal fats in the diet is one cause of **coronary heart disease**, a condition which can lead to heart attacks.

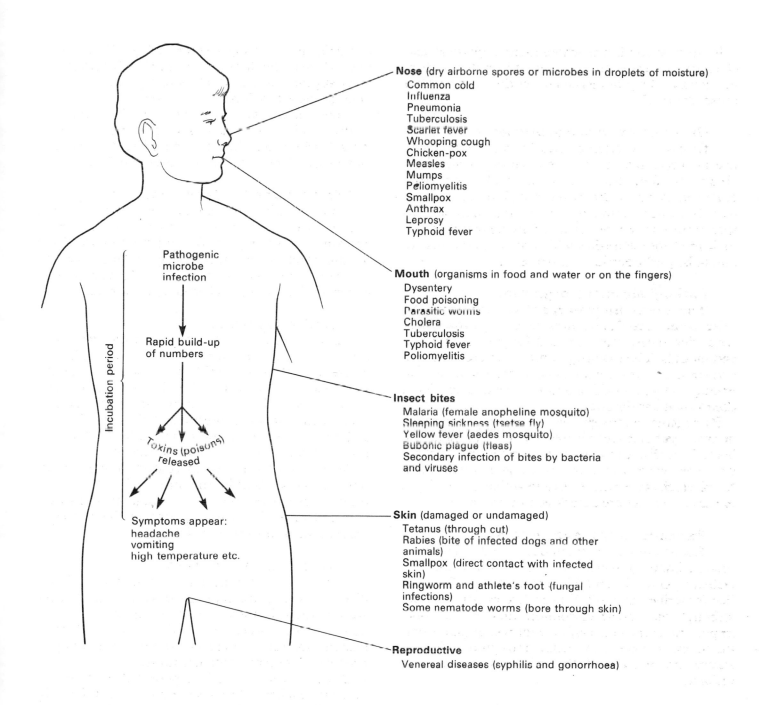

Nose (dry airborne spores or microbes in droplets of moisture)
  Common cold
  Influenza
  Pneumonia
  Tuberculosis
  Scarlet fever
  Whooping cough
  Chicken-pox
  Measles
  Mumps
  Poliomyelitis
  Smallpox
  Anthrax
  Leprosy
  Typhoid fever

Mouth (organisms in food and water or on the fingers)
  Dysentery
  Food poisoning
  Parasitic worms
  Cholera
  Tuberculosis
  Typhoid fever
  Poliomyelitis

Insect bites
  Malaria (female anopheline mosquito)
  Sleeping sickness (tsetse fly)
  Yellow fever (aedes mosquito)
  Bubonic plague (fleas)
  Secondary infection of bites by bacteria
  and viruses

Skin (damaged or undamaged)
  Tetanus (through cut)
  Rabies (bite of infected dogs and other
  animals)
  Smallpox (direct contact with infected
  skin)
  Ringworm and athlete's foot (fungal
  infections)
  Some nematode worms (bore through skin)

Reproductive
  Venereal diseases (syphilis and gonorrhoea)

Incubation period

Pathogenic
microbe
infection

Rapid build-up
of numbers

Toxins (poisons)
released

Symptoms appear:
headache
vomiting
high temperature etc.

Fig. 1. How some disease-causing organisms enter the body.

74

Some microbes which cause disease

Viruses and disease

Viruses are extremely small. A row of one million average-size viruses would be only 5 mm long. Viruses are not cells. They consist of DNA enclosed in an envelope of protein and fat. Viruses are parasites of plants and animals.

Viruses reproduce by 'hijacking a host cell' using cell materials to make hundreds of replicas of itself. The cell dies when it bursts open. It also releases the new viruses which infect other cells.

Diseases caused by viruses include chicken-pox, common colds, influenza, measles, and AIDS (Acquired Immunodeficiency Syndrome).

What is AIDS?

AIDS is a disease caused by the Human Immunodeficiency Virus (HIV). HIV weakens the body's immune system by killing T-cells, a type of white blood cell which produces antibodies. Without sufficient antibodies the body becomes defenceless against attack from many diseases which uninfected people can usually fight off without difficulty. AIDS victims are killed by chronic infections of the lungs and intestine, by cancer of the skin and bones, and by destruction of brain tissue. At present there is *no cure* for AIDS.

How people catch AIDS

The AIDS virus is transmitted when the blood or semen of an infected person comes into contact with the body fluids of another. There are two ways to catch it—by sexual intercourse with an infected person or by receiving an injection of blood from an infected person. Blood received from the Blood Transfusion Service is now safe, but sharing needles and syringes as people who inject drugs do is very risky. The most risky form of sex is any unprotected intercourse. The AIDS virus can also pass from an infected mother to her baby through the placenta.

You *cannot* catch AIDS by touching infected people and the objects they have handled, or from swimming pools which they have used. Neither can you catch it from ordinary kissing.

How to protect yourself from AIDS

1. Very simply, by avoiding casual sex. The more sexual partners you have the greater the risk of having sex with an infected person.
2. If you cannot avoid casual sex then using a condom reduces the risk of infection. But this is still risky, if the condom breaks then you have no protection.
3. Never inject drugs. Make sure that all equipment is sterilised if you have ears pierced, tatooing, or acupuncture.
4. It is best not to share razors or toothbrushes (gums can bleed when teeth are cleaned).

Bacteria and disease

Bacterial cells vary in size between 0.0005 mm and 0.005 mm in length. Like other cells they contain DNA but it is not in the form of chromosomes. The cytoplasm of a bacterium is contained within a tough cell wall and this is often covered with slime. Some bacteria can move about by means of whip-like flagella (Fig. 1).

In favourable conditions bacteria reproduce by dividing in two, a process called **fission** (Fig. 2). Under unfavourable conditions some bacteria can form thick-walled **spores** which can survive for years as dust.

Diseases caused by bacteria include boils, food poisoning, whooping cough, cholera, diphtheria, and venereal disease.

Protozoa and disease

Protozoa are complex single-celled organisms which vary enormously in size and shape. Very few cause disease in humans. *Entamoeba histolytica* is a protozoan which causes amoebic dysentery (Fig. 3A). This disease is rare in Britain but quite common in southern Europe. The amoebae can live harmlessly in the large intestine, feeding on bacteria. Under certain circumstances they become parasitic, destroying cells lining the large intestine, and forming large ulcers.

Fungi and disease

Very few fungi are parasites of humans. One of the commonest fungal diseases is called **ringworm**, because it can produce a circular swelling on the skin. Ringworm fungi also attack the scalp, and the soft skin of the groins. Another very similar fungus attacks the soft skin of the feet, especially between the toes, causing a disease called **athlete's foot**. Fungal diseases are spread by airborne spores (Fig. 4), by contact with infected people and, in the case of athlete's foot, by infected floors and mats on which people walk bare-foot.

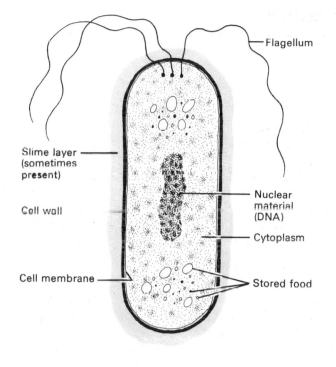

Fig. 1. **Diagram of a bacterial cell** showing most of the structures found in bacteria.

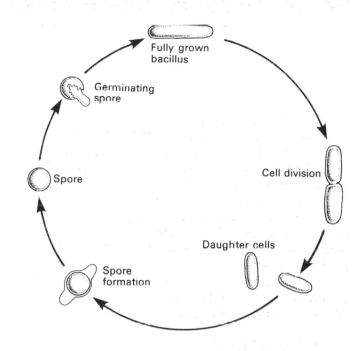

Fig. 2. **Life-cycle of a spore-forming bacterium.**

A Entamoeba histolytica × 200
(causes amoebic dysentery)

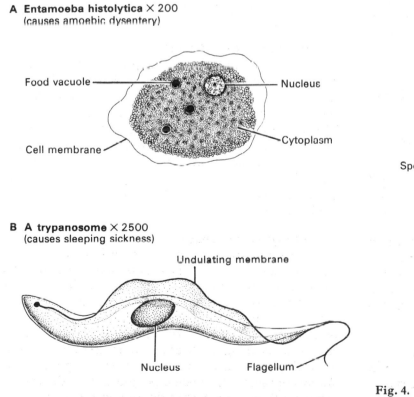

B A trypanosome × 2500
(causes sleeping sickness)

Fig. 3. **Some parasitic protozoa.**

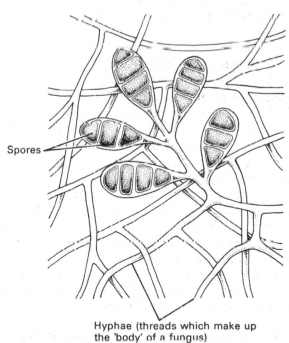

Fig. 4. **The spore-producing organs** of a microscopic fungus which causes ringworm.

75

Parasitic roundworms (Nematodes)

Roundworms are round in cross-section and have long narrow cylindrical bodies, often pointed at both ends. They vary in size from microscopic to 30 cm long. Roundworms are covered with a smooth almost transparent cuticle. Their digestive system consists of a mouth, a muscular pharynx which sucks in food, and a fairly straight gut ending at an anus (Fig. 1). They have a simple nervous system, and muscles along the body which enable them to wriggle.

Roundworms live in soil, in fresh and salt water, and as parasites in almost every type of plant and animal. About 40 species are parasites of humans, one of the commonest being a roundworm of the genus *Ancylostoma*, commonly known as the hookworm.

Human hookworms

Hookworms cause much sickness in Italy and other parts of southern Europe, in the Middle and Far East, in North and South America and in South Australia. Hookworms, and all other roundworm parasites of humans, are very rare in Britain.

Adult hookworms are between 8 and 13 mm long. They live as parasites in the intestine where they suck blood by means of a bell-shaped mouth containing several teeth. The teeth are used to cut and tear the lining of the intestine until a blood vessel is opened. The wound is then injected with chemicals which prevent the blood clotting, so that the parasite receives a plentiful supply of food. Victims with a large number of hookworms can suffer severe anaemia.

Life history

Male and female worms mate, and soon afterwards the females release fertile eggs. These are passed out of the host in its faeces.

Further development depends upon the eggs coming to rest in warm damp soil. This can happen in regions with primitive sewage disposal arrangements, or where human excreta is used to fertilize crops. The eggs hatch releasing tiny larvae which feed on bacteria in the faeces and in surrounding soil. This **first larva** grows and moults its skin to become a **second larva**, which does the same to become a **third larva**. Only the third larva can infect a new host and so it is called the **infective stage** of the life-cycle. Third larvae can bore

through thin skin on the feet, hands, and wrists of anyone who touches soil in which the larvae live. After penetrating the skin the larvae enter a blood vessel and are carried along by the bloodstream until they reach capillaries in the lungs. Here, they break out of the capillaries into the lungs and wriggle up the wind-pipe and down the gullet into the intestine. During this journey they become **fourth larvae**, which mature into adult males and females.

Control

Hookworm infection is best controlled by disposing of sewage in such a way that their eggs do not enter the soil.

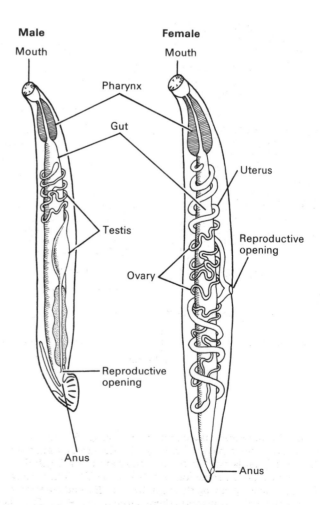

Fig. 1. Male and female hookworms. These are parasites which live in the human intestine where they suck blood.

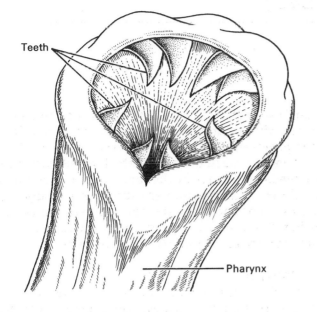

Teeth

Pharynx

Fig. 2. Mouth of a hookworm.

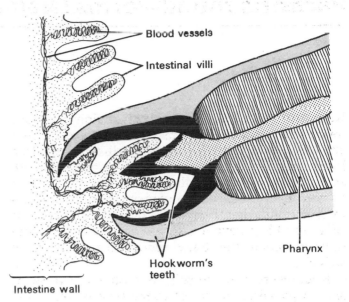

Blood vessels

Intestinal villi

Hookworm's teeth

Pharynx

Intestine wall

Fig. 3. A hookworm attached to gut wall with its teeth, sucking blood.

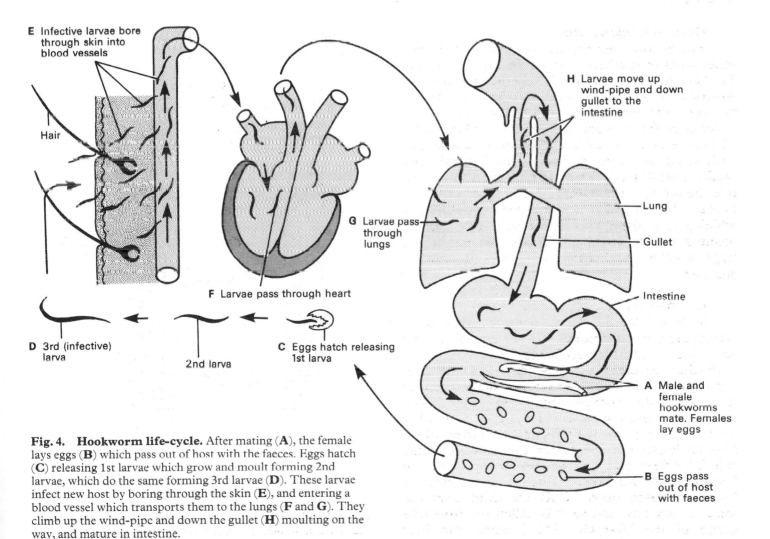

E Infective larvae bore through skin into blood vessels

Hair

F Larvae pass through heart

G Larvae pass through lungs

D 3rd (infective) larva

2nd larva

C Eggs hatch releasing 1st larva

H Larvae move up wind-pipe and down gullet to the intestine

Lung

Gullet

Intestine

A Male and female hookworms mate. Females lay eggs

B Eggs pass out of host with faeces

Fig. 4. Hookworm life-cycle. After mating (**A**), the female lays eggs (**B**) which pass out of host with the faeces. Eggs hatch (**C**) releasing 1st larvae which grow and moult forming 2nd larvae, which do the same forming 3rd larvae (**D**). These larvae infect new host by boring through the skin (**E**), and entering a blood vessel which transports them to the lungs (**F** and **G**). They climb up the wind-pipe and down the gullet (**H**) moulting on the way, and mature in intestine.

76

Tapeworms

Tapeworms belong to a group of animals called *Platyhelminths*, commonly known as flatworms because of their flattened bodies. All tapeworms are parasites, and they require at least two hosts to complete their life-cycles. The **primary host** is always a vertebrate animal which consumes the flesh of other animals. This host carries adult tapeworms in its intestine, where they mate and the females release eggs. The **secondary host** is an animal in which the eggs hatch and develop into a resting, or dormant, stage of the life-cycle called a **bladderworm** (Fig. 2D). The life-cycle is completed when the primary host eats the secondary host together with its bladderworms, which then develop into adult worms in the primary host's intestine.

Adult tapeworms

The majority of tapeworms consist of a small head called the **scolex**, from which grows a flattened tape-like body made of segments called **proglottids**. The head has suckers and sometimes hooks which attach it to the host's intestine (Fig. 1). The tape varies in length from a few centimetres to 10 metres and can consist of up to 5000 segments. Tapeworms do not have a mouth or digestive system. They absorb the host's digested food through a specialized outer skin called the **tegument**, which is resistant to the host's digestive enzymes.

Newly formed segments consist of both male and female organs; in other words they are **hermaphrodite**. But after fertilization the sex organs disappear and are replaced by several thousand microscopic eggs. At this stage the segments drop off the tape and pass out of the host's body with the faeces (Fig. 2A).

The beef tapeworm

Humans are the primary hosts of *Taenia saginata* the beef tapeworm and, as its name implies, cattle are its secondary hosts. Beef tapeworms are found throughout the world but are very rare in Britain.

If sewage disposal arrangements are primitive, tapeworm segments containing eggs can be deposited where cattle feed and drink. Segments of the beef tapeworm are unusual in that they can crawl about for a while spreading eggs on the ground.

The eggs are 0.03 mm in diameter and consist of an outer shell which protects an embryo armed with six hooks. When swallowed by a cow the egg shell is dissolved by the host's digestive juice (Fig. 2B). This releases the embryo, which uses its hooks to burrow through the cow's gut wall until it reaches a blood vessel. Embryos are carried around the cow's body until they reach muscle tissue, where they stop and develop into bladderworms about 6 mm in diameter (Fig. 2C).

Bladderworms do not harm the cow and develop no further unless the muscle (beef meat) which contains them is eaten raw, or partly cooked. If this happens live bladderworms enter the human intestine, where they quickly develop into adult tapeworms (Figs. 2 D and E).

Tapeworm infection can be prevented by efficient sewage disposal arrangements which ensure that the worm eggs cannot be eaten by cattle. Imported meat should be carefully inspected for bladderworms, which are clearly visible with the naked eye.

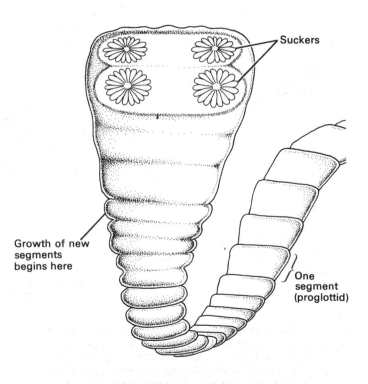

Fig. 1.   The head (Scolex) of the beef tapeworm (*Taenia saginata*). Note the four suckers which attach the head to the host's intestine wall.

Suckers

Growth of new segments begins here

One segment (proglottid)

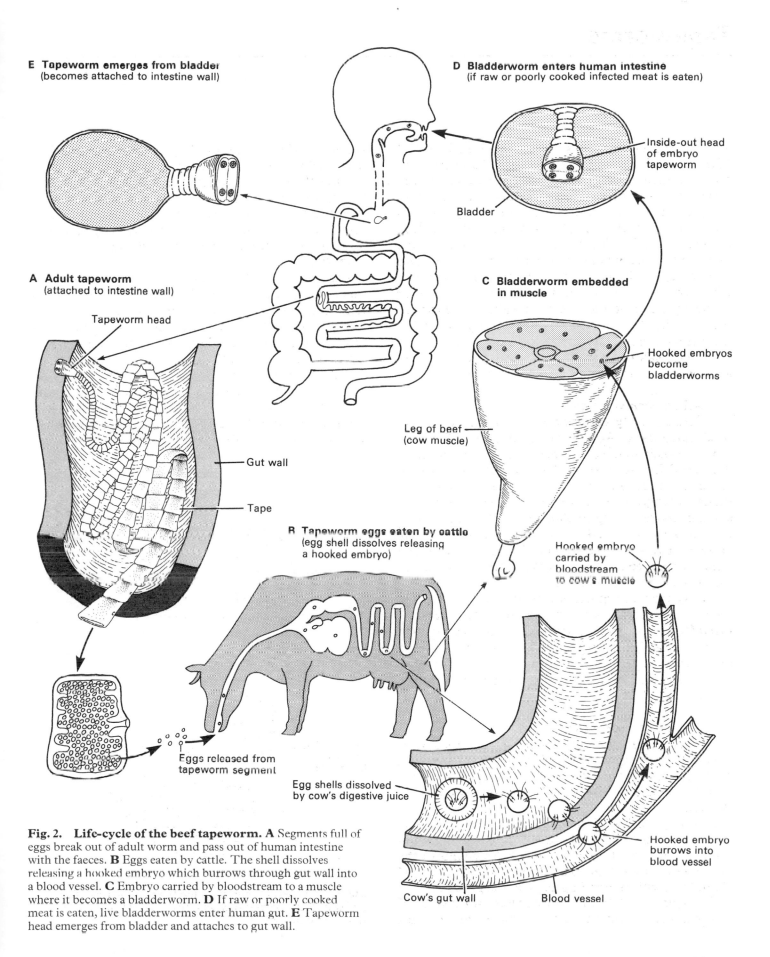

E Tapeworm emerges from bladder
(becomes attached to intestine wall)

D Bladderworm enters human intestine
(if raw or poorly cooked infected meat is eaten)

Inside-out head
of embryo
tapeworm

Bladder

A Adult tapeworm
(attached to intestine wall)

Tapeworm head

**C Bladderworm embedded
in muscle**

Gut wall

Tape

Hooked embryos
become
bladderworms

Leg of beef
(cow muscle)

B Tapeworm eggs eaten by cattle
(egg shell dissolves releasing
a hooked embryo)

Hooked embryo
carried by
bloodstream
to cow's muscle

Eggs released from
tapeworm segment

Egg shells dissolved
by cow's digestive juice

Hooked embryo
burrows into
blood vessel

Cow's gut wall

Blood vessel

Fig. 2. Life-cycle of the beef tapeworm. A Segments full of
eggs break out of adult worm and pass out of human intestine
with the faeces. **B** Eggs eaten by cattle. The shell dissolves
releasing a hooked embryo which burrows through gut wall into
a blood vessel. **C** Embryo carried by bloodstream to a muscle
where it becomes a bladderworm. **D** If raw or poorly cooked
meat is eaten, live bladderworms enter human gut. **E** Tapeworm
head emerges from bladder and attaches to gut wall.

77

Fleas, lice, and bed bugs

Fleas, lice, and bed bugs are wingless insects with piercing mouthparts which allow them to bite through skin and suck blood. The insect bite itself can be painful and may result in a swelling or rash. Furthermore, some of these insects are the carriers, or vectors, of germs which they can transfer from one host to another during feeding.

Fleas

The body of a flea is flattened from side to side (laterally compressed) which makes it easy for them to run rapidly among the hairs of their host. Each leg has a pair of claws which enable a flea to hold onto its host. The third pair of legs have powerful muscles which a flea uses to leap onto a host, and from host to host (Fig. 1). A flea can jump a distance of 30 cm or more.

Fleas live exclusively on blood but can leave their host for several days to mate and lay eggs. The female human flea *Pulex irritans* lays up to 400 eggs which she deposits in cracks, crevices, carpets, and bedding. Eggs hatch into larvae which feed on blood in the excreta of adult fleas—they do not suck blood. Larvae become pupae which hatch into adult fleas.

Humans can also be attacked by fleas whose normal hosts are dogs, cattle, pigs, rats, and mice. These have a life-cycle similar to the human flea.

Flea bites can become swollen and uncomfortable. More important, however, is the fact that fleas can transmit (i.e. are the vectors of) bubonic plague and typhus. Both of these are diseases of rats and are transmitted to humans by rat fleas. Fleas also transmit anthrax. Anthrax germs are present in the droppings of infected fleas and these can be rubbed into irritated flea-bitten skin when it is scratched.

Fleas are controlled by applying insecticide dust to floors, carpets, bedding, and rat runs. Domestic pets should also be dusted or sprayed with weak insecticide.

Lice

The bodies of lice are flattened from top to bottom (dorso-ventrally compressed), and they have claws on their legs which allow them to hold onto their host (Fig. 2A). Unlike fleas, lice complete their entire life history on their host. They lay eggs, called nits, attached to the hairs of their host (Fig. 2B). Their eggs hatch into nymphs (miniature adults) which moult several times before reaching full size.

Humans are attacked by three types of lice: two are strains of *Pediculus humanus*, one of which lives on the head and the other on the body. The third type is the crab louse *Phthirus pubis*, which lives in the pubic hairs (lower abdomen).

Lice usually spread from host to host by crawling from one to the other when they are in close contact. The bites of lice can cause skin irritation which, if scratched, can become infected with bacteria and develop into large sores. The body louse is a vector of the germs which cause endemic typhus and European relapsing fever. Infection occurs by rubbing the excreta of infected lice into their bites when the irritated skin is scratched.

Head lice can be controlled by washing the hair with a lotion called Prioderm or a medicated shampoo strong enough to kill lice and nits. A fine-toothed comb can be used to remove remaining lice and nits. Body lice are controlled by sprinkling insecticide dust inside clothing and onto bedding.

Bed bugs

The body of a bed bug is oval, and flattened like that of a louse (Fig. 3). They are called bed bugs because they usually feed on people while they sleep at night. They hide during the day in cracks and crevices. Bed bugs require fifteen minutes to complete their meal of blood and they feed every two days.

Female bed bugs lay up to 250 eggs which hatch into nymphs. In favourable conditions these reach adult size in only ten weeks, so that three or four generations may be produced in a year.

The bite of a bed bug can be very irritating and may cause lack of sleep, but there is no evidence that it transmits diseases.

Bed bugs can be controlled by spraying the floor and walls of the bedroom with liquid insecticide, and by sprinkling insecticide dust onto bedding.

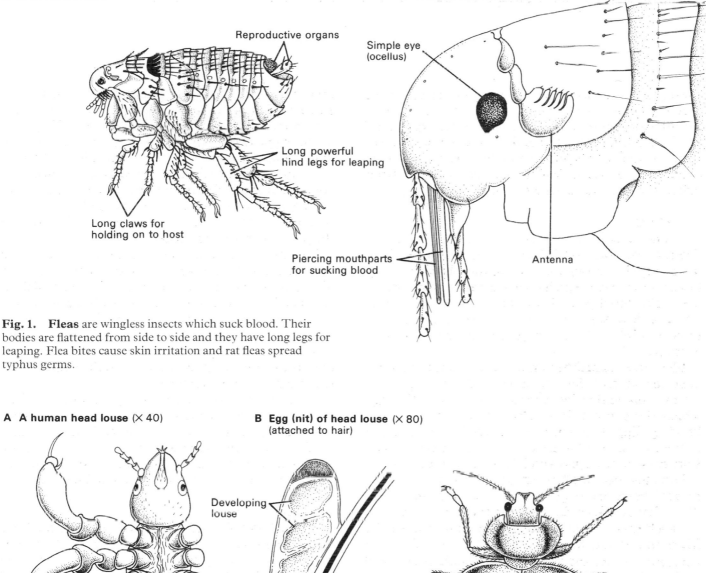

A A cat flea (×25)

Reproductive organs

Long powerful
hind legs for leaping

Long claws for
holding on to host

B Head of human flea (×150)

Simple eye
(ocellus)

Antenna

Piercing mouthparts
for sucking blood

Fig. 1. Fleas are wingless insects which suck blood. Their bodies are flattened from side to side and they have long legs for leaping. Flea bites cause skin irritation and rat fleas spread typhus germs.

A A human head louse (×40)

Long claws

B Egg (nit) of head louse (×80)
(attached to hair)

Developing
louse

Hair

(×25)

Fig. 2. Human head louse sucks blood through the scalp. Females fasten sticky eggs called nits to hairs. Lice bites can become infected with bacteria and form skin sores. Body lice spread endemic typhus.

Fig. 3. A bed bug feeds on blood but only at night. They hide during the day in cracks and crevices near the bed. Their bites can cause skin irritation but do not transmit germs.

78

Houseflies and disease

The housefly *Musca domestica* belongs to a group of insects called the *Diptera*, all of which have one pair of wings.

Feeding and mouthparts

The mouthparts of a housefly consist of a long tongue called the **proboscis** which contains a salivary channel and a food channel. Both these channels open into tiny grooves in the under surface of two flaps at the base of the proboscis (Fig. 2).

Houseflies feed on the decaying remains of dead animals, on dung, on the decaying remains of human food on rubbish tips, and on human foods and drinks containing sugar which are left uncovered.

As a fly feeds it squirts saliva containing digestive enzymes through the salivary channel of its proboscis and onto the food. The saliva partly digests the food, turning it into a liquid which is sucked up the food channel into the fly's intestine, where digestion is completed.

Life-cycle

Female houseflies lay groups of about a hundred eggs in any kind of decomposing matter (Fig. 1A). Eggs hatch into larvae commonly known as maggots, which have no head or legs (Fig. 1B). The front end has a hook which enables the maggot to crawl and burrow, and it breathes through two spiracles at the front and one at the rear. Maggots feed in the same way as an adult fly, by squirting saliva onto food.

Fully grown larvae find a dry place in which to form a pupa (Fig. 1C) from which an adult fly emerges four days later.

Houseflies spread disease

The decaying matter and dung on which houseflies feed can contain the germs which cause typhoid fever, typhus, cholera, dysentery, tuberculosis, and anthrax, together with some tapeworm and roundworm eggs. As flies feed, their feet, bodies, and digestive systems become contaminated with germs and eggs. If they are then allowed to settle on human food they spread the germs and eggs as they walk over the food, and with their saliva and droppings as they feed.

Control

Houseflies are controlled by spraying ceilings and walls with liquid insecticides, by hanging up strips of paper impregnated with insecticides, by keeping refuse in covered bins, and by keeping food in fly-proof containers.

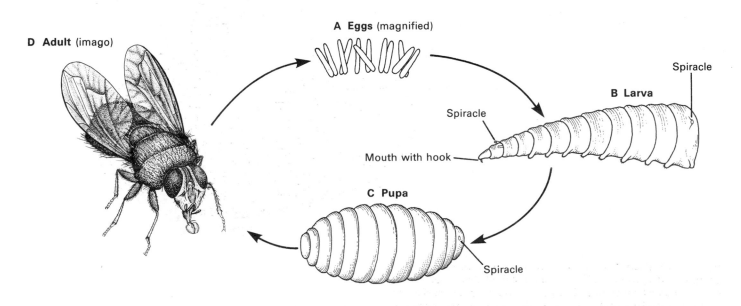

Fig. 1. Life-cycle of a housefly. A Eggs are laid on decaying matter and dung. **B** They hatch into larvae (maggots) without head or legs. The larvae become pupae (**C**) which hatch into adults four days later.

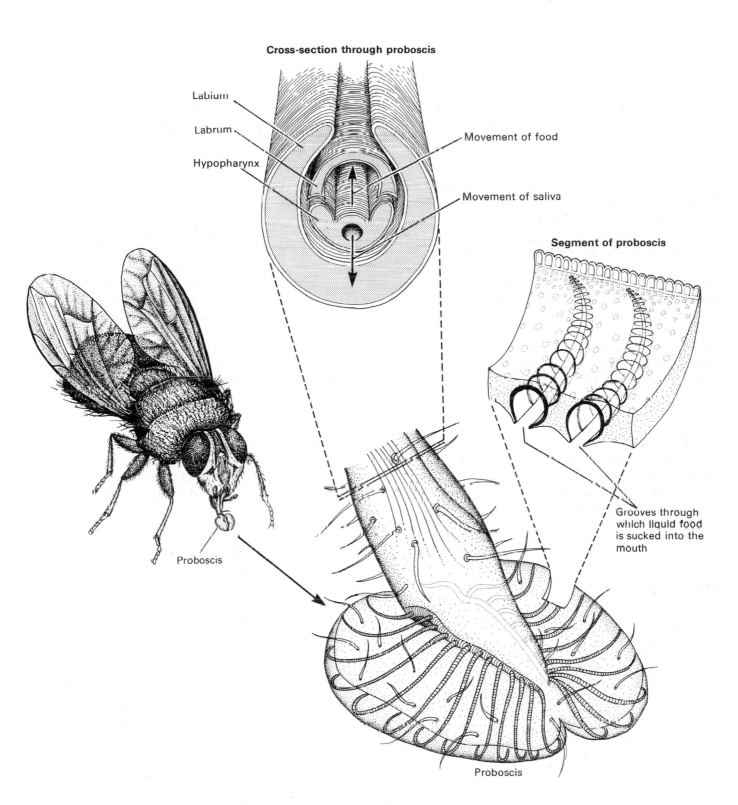

Cross-section through proboscis

Labium

Labrum

Hypopharynx

Movement of food

Movement of saliva

Segment of proboscis

Grooves through which liquid food is sucked into the mouth

Proboscis

Proboscis

Fig. 2.   Mouthparts of a housefly. A fly feeds through a tongue called a proboscis. This contains a food channel and a salivary channel which open into grooves on the under surface of two flaps at the base of the proboscis. Saliva containing digestive juice is squirted onto the food. This changes the food into a liquid which is sucked up the food channel.

79

The spread of infection

People can be infected with disease-causing organisms in many ways. The main sources of infection are: contact with infected people or with objects contaminated with germs; consumption of contaminated food and drink; bites from infected insects; and inhaling contaminated dust, or droplets of moisture breathed out by infected people.

Direct contact

Diseases spread by contact with an infected tissue of another person are said to be **contagious**. The venereal diseases gonorrhoea and syphilis can only be spread by direct physical contact because the bacteria which cause them die quickly outside the body.

Other diseases which can be spread by direct contact include fungal infections such as ringworm and athlete's foot; chicken-pox, smallpox, measles, leprosy; and boils and septic wounds.

A large number of germs can live for some time outside the body. If infected people leave such germs behind on objects which they touch, the objects then become sources of infection. This is known as spread of infection by indirect contact.

Indirect contact

The list of objects which can store germs is enormous. The most obvious are towels, coins, door handles, books, lavatory seats, a child's toys, and crockery, especially if cracked or chipped. Many of the diseases already mentioned are spread by indirect contact, especially fungal and other skin infections, and infections of the digestive and respiratory systems. It is dangerous to use the comb, towel, or bedding of an infected person. Further, since germs can be transferred from objects to the hands and then to the mouth, it is important to wash your hands before meals and after using the lavatory.

Contaminated food and drink

Food can be contaminated with disease-causing organisms by the unwashed hands of an infected person or by contact with unprotected septic wounds; by flies which have previously visited excreta and decomposing waste; by the excreta, saliva, hairs, etc. of wild rats and mice and various pets; and by contaminated water used to wash food containers or food itself. The germs of typhoid and paratyphoid fever, dysentery, and salmonella food poisoning are spread like this.

Unit 76 describes how tapeworm infections can result from eating undercooked meat containing bladderworms. Drinking water can be contaminated with faeces and urine if sewage disposal arrangements are primitive or inefficient. Cholera in particular is spread this way.

Milk and milk products such as cream can carry germs if careless dairy workers fail to wash their equipment thoroughly. Tuberculosis and brucellosis are spread to humans by the milk of infected cows.

Airborne infection

People can be infected with anthrax by breathing in dusty air containing dried spores of the bacteria which cause this disease. Airborne germs also come from the bedding, clothing, and handkerchiefs of infected people.

A sneeze can hurl 20 000 droplets of moisture containing viruses and bacteria a distance of more than four metres. Germ-laden droplets are also coughed and breathed out by infected people. Germs can float in the air for long periods and be inhaled by others. This is called **droplet infection** and is responsible for spreading colds, influenza, pneumonia, diphtheria, whooping cough, and tuberculosis.

Insects and other vectors

Houseflies gather germs on their feet, bodies, and in their digestive systems as explained in Unit 78. These germs are transferred to humans when the flies settle on food. Flies spread typhoid fever, typhus, cholera, dysentery, and tapeworm eggs.

Blood-sucking insects transfer germs amongst humans and from animals to humans. Certain mosquitoes spread the protozoa which cause malaria and the yellow fever virus, and fleas transfer bubonic plague bacteria from rats to humans.

Rabies viruses are carried by dogs, foxes, and wolves. Infected animals become vicious and unpredictable and are likely to bite humans who come near, transferring the virus in their saliva.

Healthy carriers of disease

Healthy carriers are people who carry certain disease-causing organisms in their bodies without suffering any ill-effects. Consequently they are unaware of their condition and can spread germs to other people, causing mysterious outbreaks of disease.

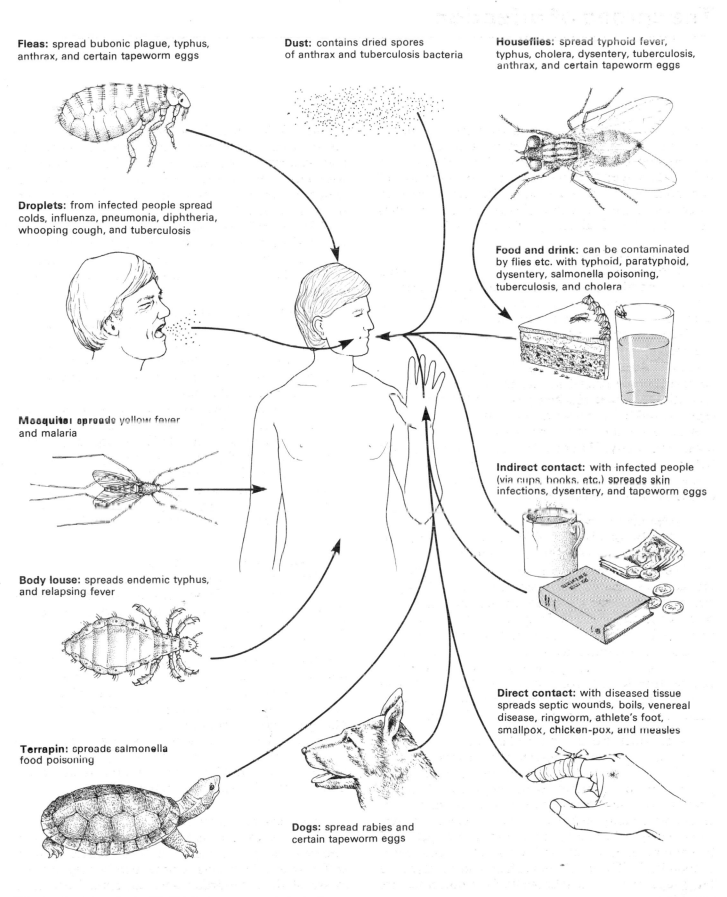

Fleas: spread bubonic plague, typhus, anthrax, and certain tapeworm eggs

Dust: contains dried spores of anthrax and tuberculosis bacteria

Houseflies: spread typhoid fever, typhus, cholera, dysentery, tuberculosis, anthrax, and certain tapeworm eggs

Droplets: from infected people spread colds, influenza, pneumonia, diphtheria, whooping cough, and tuberculosis

Food and drink: can be contaminated by flies etc. with typhoid, paratyphoid, dysentery, salmonella poisoning, tuberculosis, and cholera

Mosquito: spreads yellow fever and malaria

Indirect contact: with infected people (via cups, books, etc.) spreads skin infections, dysentery, and tapeworm eggs

Body louse: spreads endemic typhus, and relapsing fever

Direct contact: with diseased tissue spreads septic wounds, boils, venereal disease, ringworm, athlete's foot, smallpox, chicken-pox, and measles

Terrapin: spreads salmonella food poisoning

Dogs: spread rabies and certain tapeworm eggs

Fig. 1. **Some sources of infection.**

80

Prevention of infection

Good health can be maintained by keeping a good posture (Unit 17), regular exercise (Unit 18), eating a sensible diet (Unit 22), and by careful preservation of food (Unit 23).

Personal cleanliness is also very important. Care of the teeth is described in Unit 25, and Unit 44 des-cribes care of the skin and hair. Figure 1 of this Unit summarizes how health can be maintained by personal cleanliness, and Figure 2 illustrates health hazards which can result from lack of hygiene in the home.

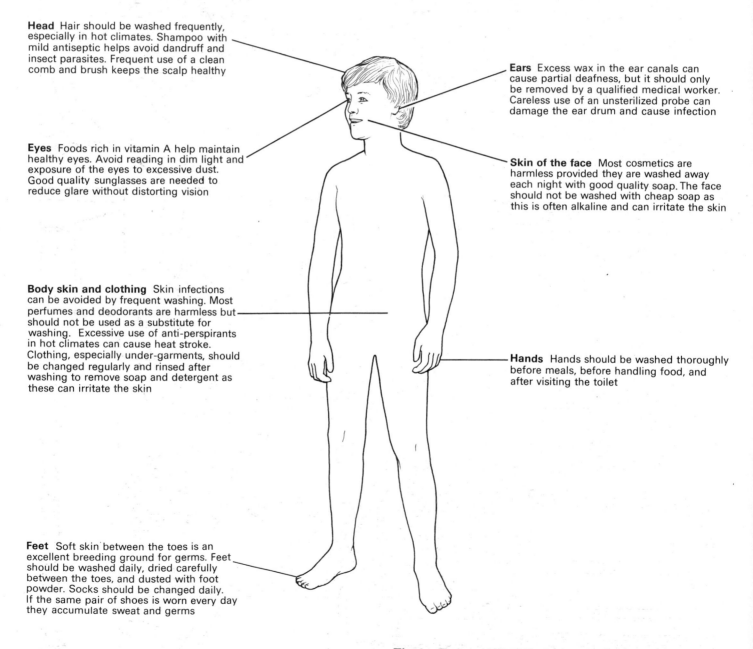

Head Hair should be washed frequently, especially in hot climates. Shampoo with mild antiseptic helps avoid dandruff and insect parasites. Frequent use of a clean comb and brush keeps the scalp healthy

Eyes Foods rich in vitamin A help maintain healthy eyes. Avoid reading in dim light and exposure of the eyes to excessive dust. Good quality sunglasses are needed to reduce glare without distorting vision

Body skin and clothing Skin infections can be avoided by frequent washing. Most perfumes and deodorants are harmless but should not be used as a substitute for washing. Excessive use of anti-perspirants in hot climates can cause heat stroke. Clothing, especially under-garments, should be changed regularly and rinsed after washing to remove soap and detergent as these can irritate the skin

Feet Soft skin between the toes is an excellent breeding ground for germs. Feet should be washed daily, dried carefully between the toes, and dusted with foot powder. Socks should be changed daily. If the same pair of shoes is worn every day they accumulate sweat and germs

Ears Excess wax in the ear canals can cause partial deafness, but it should only be removed by a qualified medical worker. Careless use of an unsterilized probe can damage the ear drum and cause infection

Skin of the face Most cosmetics are harmless provided they are washed away each night with good quality soap. The face should not be washed with cheap soap as this is often alkaline and can irritate the skin

Hands Hands should be washed thoroughly before meals, before handling food, and after visiting the toilet

Fig. 1. **Personal cleanliness** is a very important part of the fight against infection.

Fig. 2. Health hazards in the home. Make a list of the hazards to health illustrated in this drawing.

81

Pure water supplies

On average one person uses 160 litres (35 gallons) of water every day. Where does this water come from and how is it made pure enough to drink?

The main sources of water are: boreholes (deep wells) which extract water from water-filled rocks called **aquifers** (Fig. 1A); reservoirs, which collect rainwater from a region called a **catchment area** (Fig. 1B); and rivers from which water is **abstracted** (Fig. 1C).

Purification of water

The purest water comes from deep boreholes. Reservoir water is fairly pure but river water often contains harmful bacteria and chemicals. These are removed in the following ways.

Screening Metal grids are placed at the points where water is taken from rivers and reservoirs to keep out weeds and other floating debris (Fig. 1C).

Sedimentation Water is passed through tanks where large particles settle to the bottom forming a sediment, or sludge, which is piped away.

Coagulation Chemicals such as alum or ferrous sulphate are added to water. These make particles in the water stick together (coagulate) forming larger particles which are then filtered out.

Filtration Water is filtered through a layer of sand or other material which removes suspended particles.

Chlorination Chlorine is added to water to kill bacteria. The water is then stored in covered reservoirs or water towers until it is used.

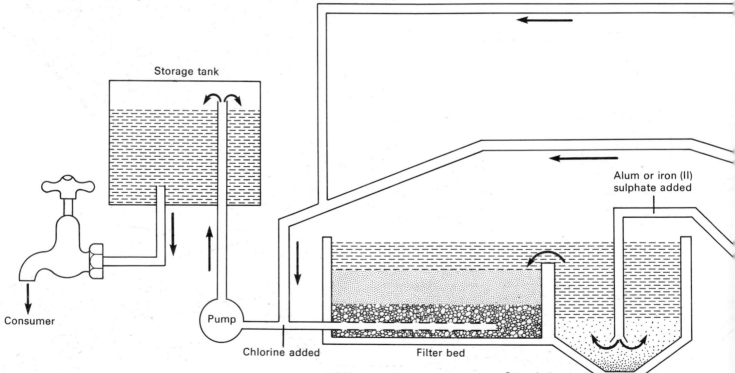

Fig. 1. Sources of water and its treatment. Screening removes floating debris. Sedimentation allows large particles to settle out. Chemicals are added to make particles stick together (coagulate) so that they are easily filtered. Chlorine kills bacteria in the water.

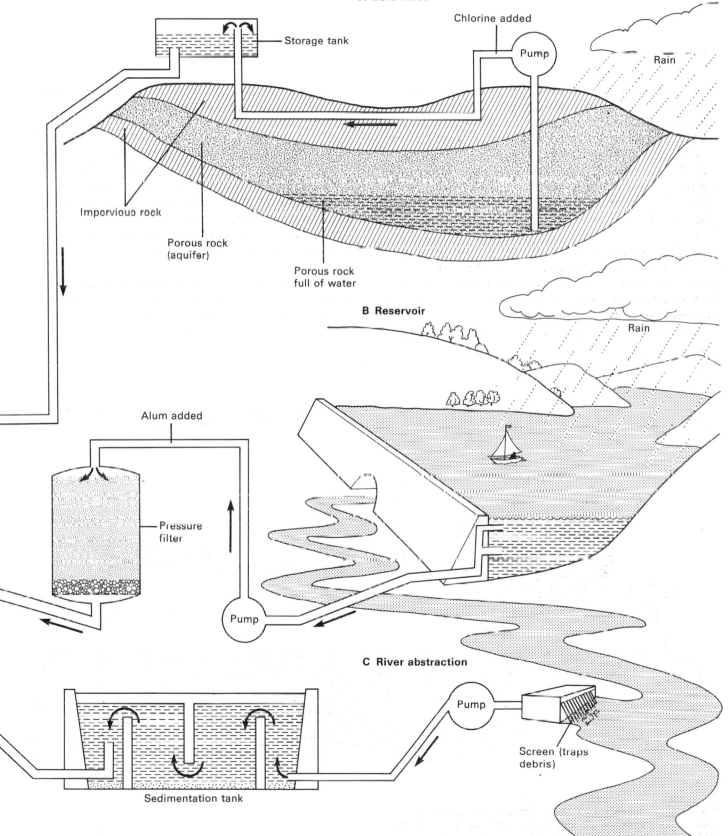

A Bore holes

Chlorine added

Storage tank

Pump

Rain

Impervious rock

Porous rock
(aquifer)

Porous rock
full of water

B Reservoir

Rain

Alum added

Pressure
filter

Pump

C River abstraction

Pump

Screen (traps
debris)

Sedimentation tank

82

Disposal of sewage and refuse

Sewage consists of human faeces and urine diluted with some rainwater, and water from kitchens and washrooms of houses, shops, factories, etc. Refuse, or garbage, consists mainly of paper, empty cans, bottles, and unused food.

Sewage and refuse can contain germs and eggs of parasitic worms. Consequently it is very important that they are disposed of so that they do not contaminate drinking water, and so that flies, rats, and other vectors cannot find them and spread infection.

Sewage disposal

In camps and isolated houses sewage can be disposed of in a hole filled with enough soil to keep out flies and rats. Touring caravans and aircraft often use chemical closets to dispose of sewage. The chemicals added to the sewage remove unpleasant smells, kill bacteria, and liquefy faeces.

Isolated houses with piped water often make use of septic tanks to dispose of sewage. Faeces and urine are flushed out of the lavatory (Fig. 1) into a large concrete-lined tank some distance from the house. Here, solid matter in the sewage is consumed by bacteria and protozoa, which break it down into harmless substances which are removed about once a year.

City sewage is flushed into large sewers where it is carried by a flow of water to a sewage treatment works (Figs. 2 and 3). Here, solids and bacteria are removed leaving an **effluent** which is harmless enough to be released into rivers, lakes, or the sea, without causing pollution.

Sewage treatment

Screening   Metal grids strain out large solid objects like sticks, paper, rags, etc. (Fig. 3A). These are removed by conveyor belts and burned or buried.

Grit tanks   Screened sewage passes into tanks where its rate of flow is reduced so that grit, stones, and other heavy solids sink to the bottom (Fig. 3B).

Sedimentation tanks   Sewage is next channelled into tanks where it remains still long enough for lighter solids such as faeces to settle to the bottom (Fig. 3C). Fluid, called **liquor**, at the top of the tank is drained off for further treatment. Sludge at the bottom of the tank is dried to form fertilizer, or fermented with bacteria to make methane gas for domestic and commercial use.

Biological filtration   Liquor from sedimentation tanks contains dissolved organic materials which can be removed 'biologically' using micro-organisms. The liquor is sprinkled into circular tanks filled with clinker, stones, and coke (Fig. 3D). This material is quickly covered with a film of bacteria and protozoa which consume the organic materials, changing them into carbon dioxide and other harmless waste. Air spaces between the clinker allow oxygen to reach the micro-organisms.

Filtered liquid is passed through a final sedimentation tank where clinker, dust, and bacteria settle out and it is then ready to be discharged.

Refuse disposal

Refuse should be collected in bins with well-fitting lids to keep out the flies, or in tough paper sacks. It is then spread onto conveyor belts and metal, glass, and paper are removed for re-use (recycling).

Inflammable rubbish is burnt in an incinerator. The remaining rubbish is spread onto waste ground, sprayed with strong disinfectant, allowed to settle, and then covered with a layer of soil.

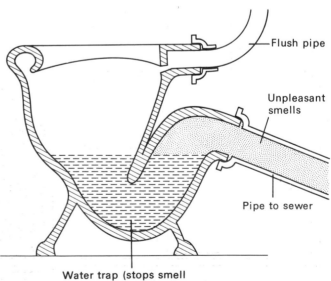

Labels: Flush pipe — Unpleasant smells — Pipe to sewer — Water trap (stops smell from sewer entering house)

Fig. 1.   Diagram of a lavatory pan or water closet (W.C.). Water is flushed around the rim, down the sides, and through the U-bend. Water retained in the U-bend seals off smells from the sewer.

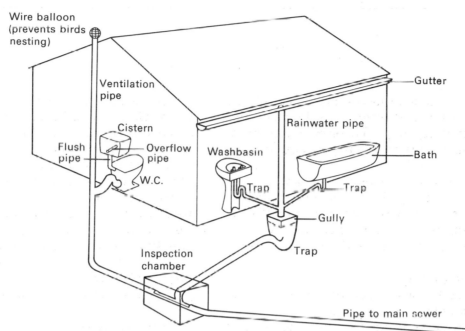

Fig. 2. Waste pipes and drains from a house.

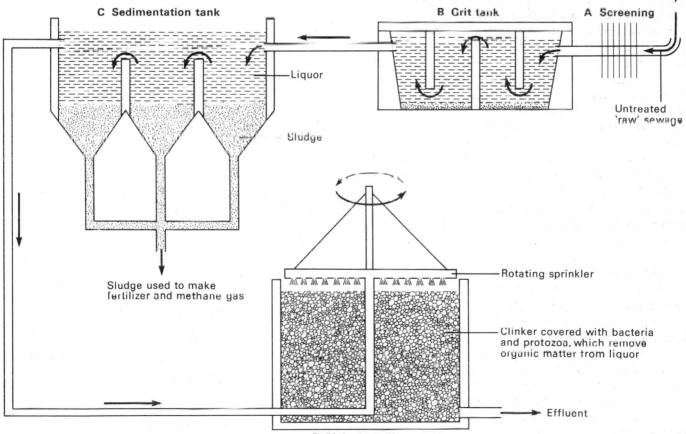

Fig. 3. Treatment of sewage. Screening (**A**) removes large objects, and small heavy particles settle out in the grit tanks (**B**). Sedimentation (**C**) separates a sludge (which is made into fertilizer and methane) from a liquor (which is cleaned by micro-organisms in the biological filter (**D**)).

165

83

Safety in and around the home

In Britain alone six and a half thousand people are killed and a million and a half are injured each year by accidents in their own homes. 44 per cent of these accidents involve children. This means that the home is a dangerous place, especially for the young.

People injure or kill themselves at home for many reasons, but the following are among the most important.

Carelessness and stupidity

People usually feel safe at home and this can lead to carelessness and downright stupidity. They try to mend an electric iron or toaster while it is still plugged in; they change light bulbs or try to reach wall cupboards while standing on a rickety chair; they leave razor-sharp knives or dangerous pills which look like sweets where they can be found by young children; they put bleach, paint stripper, or other corrosive liquids into glass bottles which are easily broken; they leave a pan of boiling liquid with its handle within reach of a curious child, and so on.

Bad design

The Royal Society for the Prevention of Accidents is trying to discover which household articles are involved in the greatest number of accidents. Often the article proves dangerous because it is badly designed.

Toys, for example, may have sharp points or edges and may be covered with poisonous paint; pans with heavy handles may tip over at the slightest touch, and electrical equipment may have loose wires which suddenly make it 'live' and able to give a fatal shock.

Dangers outside the home

Apart from accidents in the street or at work there are numerous ways in which people can be killed or injured outside their homes. A few of them are illustrated in Figure 1.

It is most important to remember that homes and their surroundings are dangerous, and to be especially vigilant when young children are present.

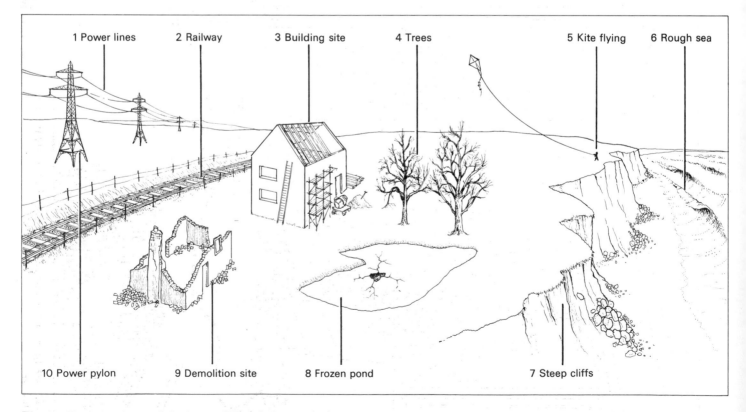

1 Power lines 2 Railway 3 Building site 4 Trees 5 Kite flying 6 Rough sea

10 Power pylon 9 Demolition site 8 Frozen pond 7 Steep cliffs

Fig. 1. Dangers around the home. What dangers exist in the numbered parts of this scene?

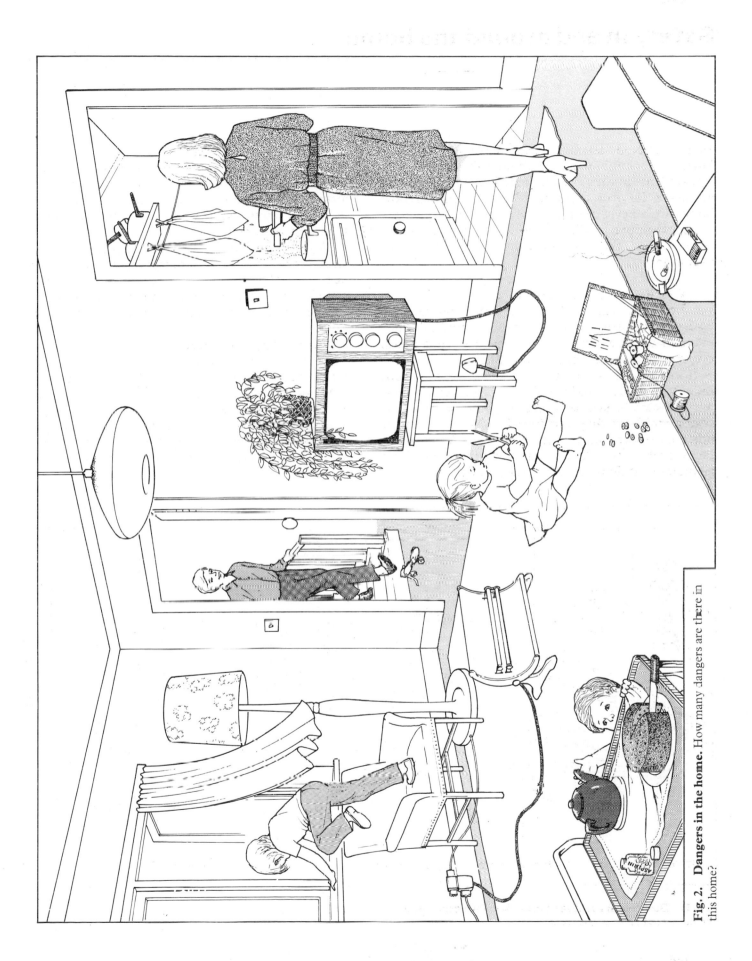

Fig. 2. Dangers in the home. How many dangers are there in this home?

167

84

Ventilation, heating, and lighting

Good health, resistance to infection, and mental alertness depend, to a certain extent, upon comfortable surroundings at home and at work. Homes and places of work cannot be comfortable unless they are properly ventilated and heated, and provided with adequate lighting.

Ventilation

The air out-of-doors is: purified by ultraviolet rays from the sun which kill bacteria; washed free of dust by rain; and charged with oxygen by photosynthesis in green plants.

Air inside a building, however, is not kept clean and fresh by these natural processes. A poorly ventilated room containing people soon becomes unpleasantly warm, smelly, and humid (stuffy), and contains a large number of germs breathed out by its occupants. The aim of ventilation is to remove this stale air and replace it with fresh air from outside. Ideally, this should be done without causing fast-moving air currents (draughts) as these can result in rapid cooling of the body (hypothermia) which is both unpleasant and harmful to health, especially in old people.

The simplest way of ventilating a room is to open windows (Fig. 1A). Alternatively, an extractor fan can be used to suck air out of a room such as a kitchen (Fig. 1B). Complex air-conditioning equipment is available which sucks in air from outside, filters out any dust, cools or warms it to the desired temperature, and adjusts its humidity.

Heating and insulation

Heat can be obtained from open fires burning wood, coal, or smokeless fuel; from various types of gas, electric, or kerosene (paraffin) fires; or from central heating systems in which heated water or air is pumped around a building through pipes and radiators or air ducts (Fig. 2).

Central heating systems and electric fires heat air without adding moisture, thereby reducing humidity. The resulting dry air can cause discomfort to people with lung disorders like bronchitis. As long as care is taken over safety, humidity can be increased by placing wet cloths or open containers of water on radiators or near electric fires.

Excessive heat loss from a building wastes both money and energy. Figure 3 illustrates ways of keeping heat loss to a minimum.

Lighting

Adequate lighting indoors is essential for safety in places such as stairs, and to avoid eye strain during reading and writing.

Electric lighting from filament lamps and fluorescent tubes is the most acceptable alternative to natural light from the sun. Lights should be arranged to avoid intense shadows and brightly lit areas, as the contrast between the two causes eye strain. Lighting can be improved by using pale-coloured paints and wall-papers to reflect light from walls and ceilings.

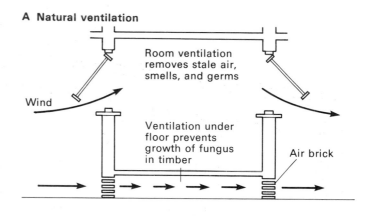

A Natural ventilation

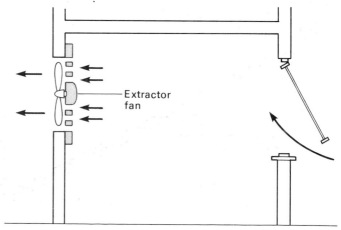

B Mechanical ventilation

Fig. 1. Ventilation removes stale, germ-laden air from a room. This can be done by opening windows (**A**), or by using fans which extract air from, or force air into, a room (**B**).

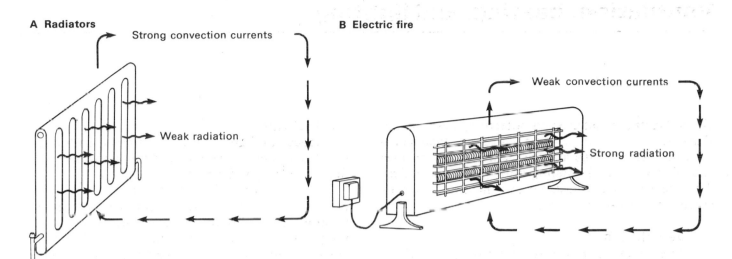

A Radiators

Strong convection currents

Weak radiation

B Electric fire

Weak convection currents

Strong radiation

Fig. 2. **Heat** is distributed by convection (air currents) and by radiation (heat rays).

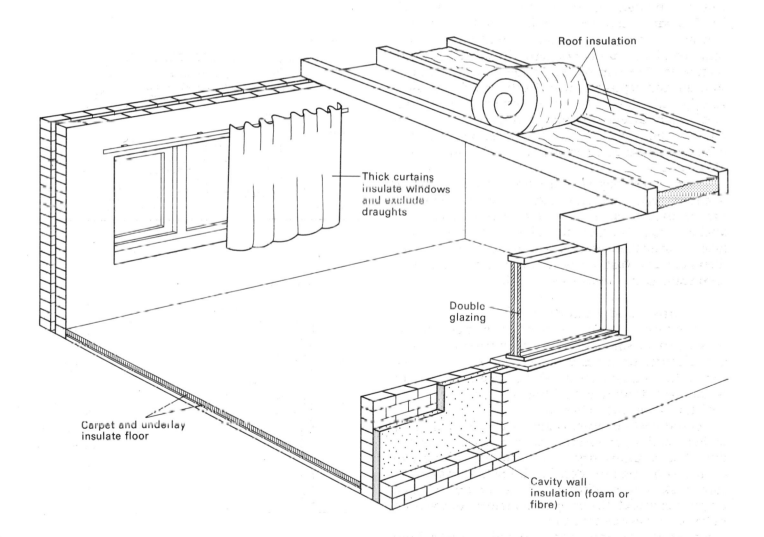

Roof insulation

Thick curtains insulate windows and exclude draughts

Double glazing

Carpet and underlay insulate floor

Cavity wall insulation (foam or fibre)

Fig. 3. **Insulation** reduces the rate at which heat is lost from a building.

85

Pollution

A substance becomes a **pollutant** when its presence in air, water, or soil harms living things. Pollutants harm humans in several ways. In high concentrations they can cause ill health and even death. They can spread through food chains (described in Unit 19) destroying plants and animals and endangering human food supplies like fish, and they can cause dirt and unpleasant smells.

Air pollution

The main sources of air pollution are the burning of coal and oil in houses and factories, and in the engines of cars, buses, aeroplanes, etc.

Smoke produced by burning contains small particles of dust which are mainly carbon. This dust blackens the walls of buildings and settles on the leaves of plants, limiting photosynthesis by cutting out light, and limiting transpiration by blocking stomata. Smoke also contains sulphur dioxide. This gas reacts with water vapour in air forming sulphuric acid which damages the stonework of buildings, the leaves of plants, and people's lungs.

Garden bonfires can also be a source of dangerous pollution. If household rubbish including plastic and polystyrene is added to the fire, its smoke will contain up to 300 times more cancer-producing chemicals than cigarette smoke, as well as cyanide, lead, dioxin, and other poisonous chemicals.

Petrol and diesel engines release fumes containing oxides of nitrogen and lead compounds. Once lead enters the body it cannot be removed by the excretory system. It collects in the body eventually causing damage, especially to the brain.

Water pollution

The main sources of water pollution are sewage from houses and farms, chemical waste from industry and agriculture, and spilled oil.

Sewage can be made harmless as described in Unit 82. But in many countries population growth has overloaded sewage treatment works and untreated sewage is released into rivers and the sea. Sewage is decomposed by bacteria in water, but in lakes and slow-moving rivers this process uses up oxygen so quickly that fish, insects, tadpoles, etc., are killed.

Industrial waste often contains very poisonous, long-lasting pollutants such as compounds of cyanide, lead, mercury, and copper. These chemicals are dangerous even in small quantities, because when they are released into streams and rivers they accumulate in fish and other aquatic creatures. They then spread through food chains to water birds, otters, and sometimes humans.

In many modern farms poultry, cattle, and pigs are kept in buildings and there is no other land on which to use the manure which they produce. The manure is discharged into local streams and rivers, where it decomposes and reduces oxygen levels in the same way as untreated domestic sewage. Other pollutants resulting from modern farming methods include chemical sprays which kill insect pests and fungi that attacks crop plants. If these chemicals enter rivers and ponds they can spread throughout food chains in the same way as industrial waste.

Radiation

Radiation such as X-rays and beta and gamma rays can cause various types of cancer, a blood disorder called leukaemia, and damage to the sperms and ova resulting in deformed babies.

Natural radiation comes from outer space in the form of cosmic rays, and artificial radiation comes from certain medical and industrial processes. Little if any harm comes from these sources; but there is increasing concern about radiation from the testing of nuclear weapons and from nuclear power stations.

As oil and coal supplies are used up the need to use nuclear power to generate electricity will increase. There is always a risk that radiation will escape from a nuclear power station, but the main problem is how to dispose of the radioactive waste produced. One method is to encase it in glass or concrete and dump it in the sea or bury it deep in the earth. However, the waste remains radioactive for long periods, sometimes thousands of years.

Noise

Noise from cars, motorcycles, aeroplanes, dogs, children, radios, televisions, etc. is sometimes described as pollution because it can cause mental and physical harm.

Prolonged loud noise from sources such as industrial machinery and discotheque music can damage the inner ear and cause partial deafness. Noises can also cause sleeplessness and mental depression. In some cases it is possible to take legal action against people who disturb others by creating noise over a long period.

A   Car exhaust fumes contain oxides of nitrogen and lead.

B   Factory smoke contains carbon particles and sulphur dioxide gas.

C   Thoughtless dumping of rubbish is offensive to the eye and endangers wild life.

D   Detergents released from factories and homes destroy fish and other aquatic creatures.

E   Noise is also a form of pollution.

F   Oil washed ashore from a damaged tanker.

Smoking and health

Over the past 100 years improvements to housing, sanitation, and medical services in Britain have resulted in a more healthy population. Nowadays some of the major hazards to health are self-inflicted. These hazards include smoking, which is described in this Unit, and drinking and drug taking which are described in Unit 87.

Smoking

There is now an enormous amount of evidence, summarized in Figure 1, that the following diseases occur more often in smokers than non-smokers: lung cancer; emphysema (thinning and weakening of lung tissue); cancer of the mouth, throat, larynx, gullet, bladder, and pancreas; coronary thrombosis (blockage of arteries to the heart); angina pectoris (pain due to narrowing of arteries to the heart); and chronic bronchitis with phlegm. In addition, smoking appears to delay the healing of stomach ulcers; it reduces the senses of smell and taste; slows down reflexes (making smokers more prone to accidents); and gives the breath, clothes, and homes of smokers an unpleasant smell.

Recent research shows that smokers endanger the health of non-smokers. Pregnant women smokers tend to have smaller babies than non-smokers, and their babies are more likely to be born dead or die a few days after birth. Furthermore, the children of smoking parents have more lung infections in the first years of life than the children of non-smokers. During one hour in a smoky room a non-smoker can inhale as much cancer-causing substance as someone smoking fifteen filter-tip cigarettes.

Illness related to smoking is very costly. In Britain it results in the loss of fifty million working days a year, treatment costs several hundred thousand pounds a day, and it causes the death of over a thousand people a week. However, those who give up smoking greatly reduce their chances of developing the diseases listed above (Fig. 1 B and E).

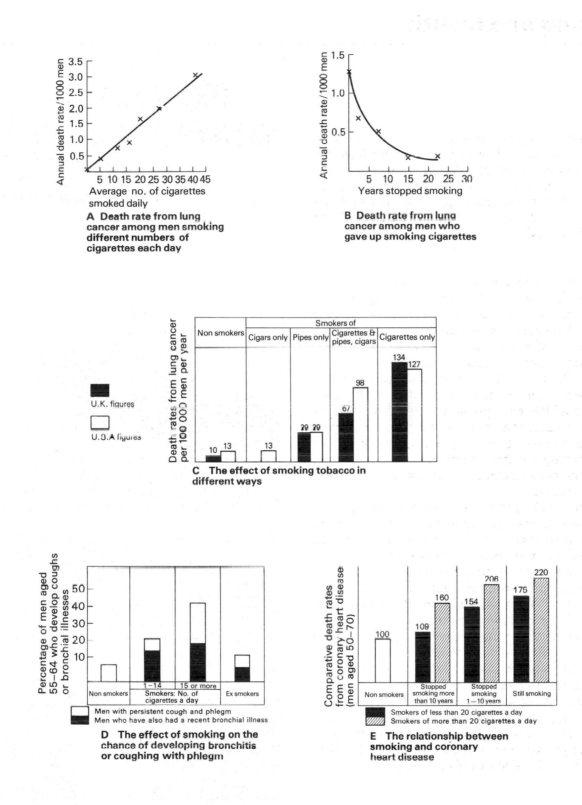

A Death rate from lung cancer among men smoking different numbers of cigarettes each day

B Death rate from lung cancer among men who gave up smoking cigarettes

C The effect of smoking tobacco in different ways

U.K. figures

U.S.A figures

D The effect of smoking on the chance of developing bronchitis or coughing with phlegm

Men with persistent cough and phlegm
Men who have also had a recent bronchial illness

E The relationship between smoking and coronary heart disease

Smokers of less than 20 cigarettes a day
Smokers of more than 20 cigarettes a day

Fig. 1. The effects of smoking on health (figures taken from 'Mortality in relation to smoking: ten years' observations by British doctors', by R. Doll and A. B. Hill, *British Medical Journal*, Vol. 1, 1964). The statistics in these graphs refer to men because in the past men have tended to smoke more heavily than women. But women who smoke are liable to contract the same illnesses as men.

Alcoholism

Alcoholics do not drink simply for pleasure, but because they feel that they cannot face life's problems without alcohol. In other words they have become **dependent** upon alcohol. People can become dependent upon alcohol without realizing it. At first pleasing effects are produced by one or two drinks, but soon larger and larger amounts are required to give these effects, until self-control is lost.

More road accidents result from heavy drinking than from any other cause. This is because alcohol slows down reflexes, interferes with concentration and distance judgement, and increases recklessness.

Heavy drinking can cause cancer of the mouth, gullet, stomach, and liver. It upsets the digestion and reduces blood cell formation, causing anaemia. Alcohol causes shrinkage of the brain, reducing the powers of abstract reasoning, and it destroys liver cells causing this organ to store abnormally large amounts of fat. In severe cases, alcoholics suffer numbness and paralysis of the limbs.

Some alcoholics suffer a disorder called **delirium tremens** when forced to stop drinking. Vomiting occurs and the whole body begins to shake. This is followed by grotesque and often terrifying visions (hallucinations).

Drug taking

A drug can be defined as a chemical which affects the mind. Only the so-called 'hard' drugs are addictive. These include the opiates: opium, morphine, and heroin. A drug addict comes to depend upon a drug so that life is no longer bearable without it. Addiction occurs because, like alcohol, more and more of a drug is required to produce its desired effects. Moreover, if the drug supply is suddenly cut off an addict suffers **withdrawal symptoms**. These are extremely unpleasant and can be fatal.

Marijuana and hashish These drugs are produced from the plant Indian hemp. Visible effects of the drugs include reddening of the eyes caused by dilation of blood vessels, and enlargement of the pupils. Effects on the mind vary. In general sensory awareness is increased and ideas flow more quickly. These drugs are not addictive and there is no evidence that they harm the body. There is a danger, however, that the drug taker's sense of judgement will be distorted so that reckless or foolish behaviour can result.

LSD (lysergic acid diethylamide) LSD is an example of an **hallucinogenic drug**, so-called because it gives rise to dramatic hallucinations.

The effects of an LSD 'trip' are unpredictable and seem to depend upon the mood of the user immediately before taking the drug. It could, for instance, intensify awareness and perception to the point at which the user undergoes mystical experiences, or it may intensify a depressed, fearful, or angry mood with horrifying results.

Another problem with LSD is that very small quantities have powerful effects, and since illegally produced drugs vary in quality drug takers can never be sure how much they are taking. An overdose of LSD can result in insanity or death.

Opiates (opium, morphine, and heroin) These drugs are extracted from the seed capsules of the opium poppy. Opiates are medically important as pain-killers. But drug takers use them because they give rise to feelings of well-being, contentment, and power. Unfortunately when the effects wear off the taker becomes anxious and depressed and is tempted to take another dose to restore a good mood, a course which can lead to addiction.

Morphine, heroin, and other drugs which are often injected involve the risk of infection from dirty hypodermic needles.

Barbiturates and amphetamines Barbiturates are used medically to relieve anxiety and as sleeping pills. But they are dangerous because the dosage must be continually increased to be effective.

Barbiturates are often used in combination with amphetamines and other stimulants. These replace the sleepiness induced by barbiturates with a feeling of mental alertness. Amphetamines harm the health by reducing appetite, by causing sleeplessness, and by reducing the body's ability to fight infection.

Possession of the drugs described above is illegal unless they are being used under medical supervision. The illegal use or sale of these drugs can result in heavy fines and prison sentences.

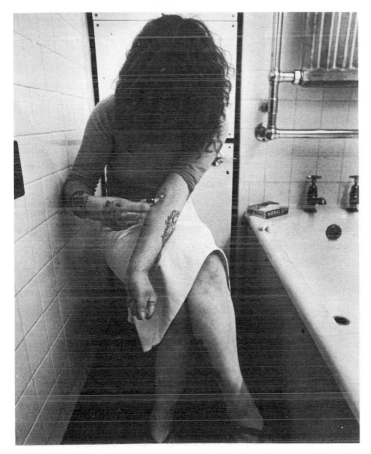

Apart from the dangers of addiction, drugs which are injected involve the risk of infection from dirty hypodermic needles.

Heavy drinking can lead to both mental and physical deterioration. Alcoholics can lose the will to care for themselves properly.

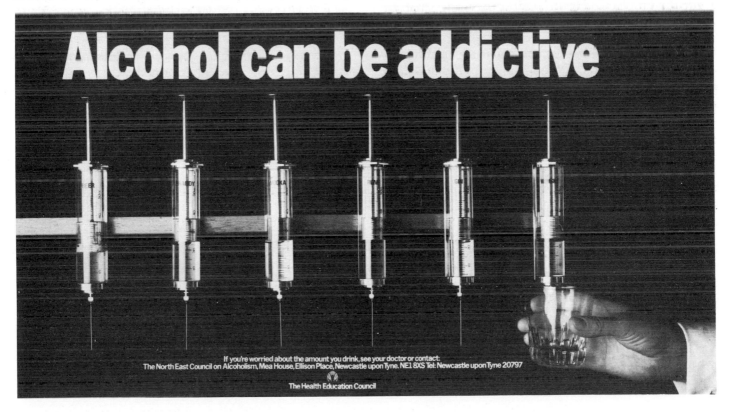

If you're worried about the amount you drink, see your doctor or contact:
The North East Council on Alcoholism, Mea House, Ellison Place, Newcastle upon Tyne. NE1 8XS Tel: Newcastle upon Tyne 20797

The Health Education Council

88

The National Health Service

The British National Health Service began operating in July 1948. Its aim is to provide 'a comprehensive medical service to secure improvement in the physical and mental health of the people, and the prevention, diagnosis and treatment of illness'.

The National Health Service is provided by the Department of Health and Social Security and led by an elected Member of Parliament, the Secretary of State, who is helped by a team of civil servants. The Service is paid for mainly out of taxes together with contributions paid by employers, and by employees according to their income, plus separate charges for prescriptions and dental and optical services.

The National Health Service provides **Personal Health Services** concerned with the health of individual patients, and **Environmental Health Services** concerned mainly with preventing rather than curing illness.

Personal Health Services

Included under this heading are services provided by hospitals and General Practitioners.

Hospital services Most British hospitals are controlled by Local Health Authorities, and these are controlled by Regional Health Authorities. Hospitals look after both in-patients and out-patients. In addition to wards for in-patients, hospitals have X-ray departments and laboratories to help in the diagnosis of disease. Cardiology units care for heart ailments, radiology units use radiant energy to treat disease, and blood transfusion units provide blood. Some hospitals fit appliances such as artificial limbs.

General Practitioner services General practitioners include doctors, dentists, opticians, chiropodists, and pharmaceutical chemists.

Environmental Health Services

These services are controlled by Local Authorities and are paid for by grants from the Department of Health and Social Security and from rates paid by property owners. Local Authorities look after the following aspects of health.

Ambulance services This is a 24-hour service. It is connected with the 999 telephone call system for emergencies. 999 callers should give clear information about the place and nature of the accident or illness.

Vaccination and immunization Immunization gives a person the ability to resist infection, and is usually achieved by injecting a vaccine (described in Unit 91). It is advisable to immunize children against poliomyelitis, measles, diphtheria, tetanus, and whooping cough early in life.

Maternity services Maternity clinics are provided to help pregnant women in many ways. They test women for signs of venereal disease, German measles, high or low blood pressure, and rhesus incompatibility (explained in Unit 35), all of which can harm the unborn child. An unborn baby's growth rate can be measured and there are ways of detecting physical deformities and hereditary diseases before birth. Regular attendance at these clinics during pregnancy greatly reduces the risk of complications at childbirth which could lead to the death or illness of the mother or her newborn child.

Midwife service Midwives are trained to care for and advise mothers during and after pregnancy and childbirth.

Home-help service Home-helps are made available to do the shopping, cooking, washing, and cleaning for the sick and aged.

Health visitors These are nurses trained to visit homes and give advice on the care of children, expectant and nursing mothers, the elderly and sick people, and to help in preventing the spread of infection.

Home nurses These are nurses who visit homes to care for the sick of all ages under the direction of a doctor.

Mental health service This service provides trained staff to care for the mentally ill and the mentally subnormal.

Local Authorities are also responsible for maintaining a clean environment by providing equipment and staff for sewage and refuse disposal, by clearing or repairing slum houses and, where necessary, providing new houses. They also detect and remove sources of air and water pollution, and test food and drugs to ensure their purity.

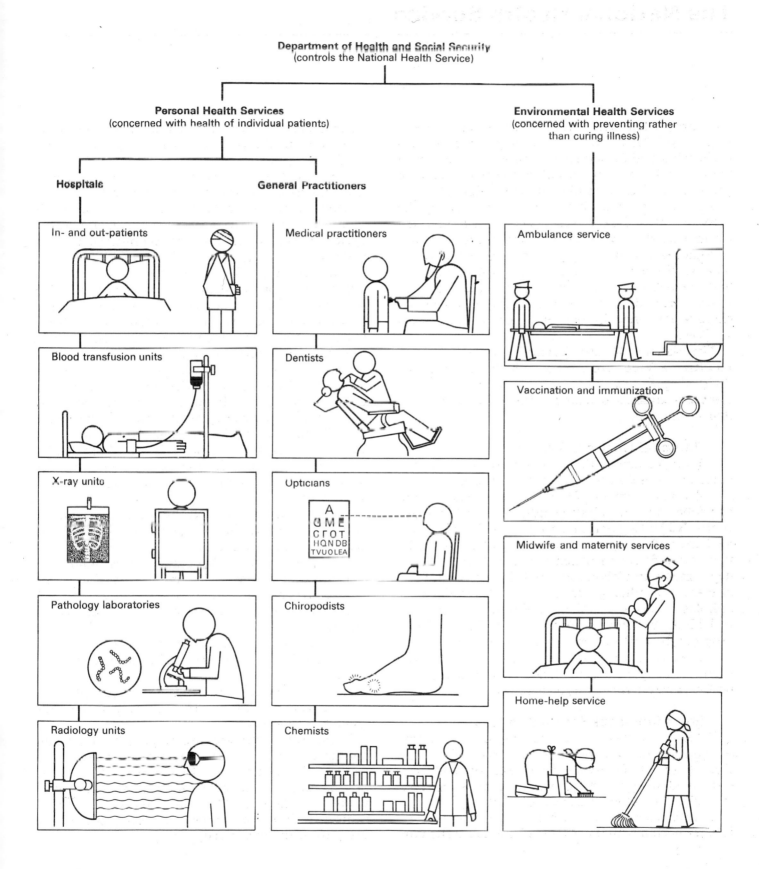

Fig. 1. Some functions of the National Health Service.
The Service is controlled by the Department of Health and
Social Security. Personal Health Services are controlled directly
by this Department. Environmental Health Services are the
responsibility of Local Authorities.

89

Antiseptic surgery and medical drugs

The work of a French biologist, Louis Pasteur (1822–95), and a German doctor, Robert Koch (1843–1910), proved that micro-organisms cause disease. Pasteur also discovered that micro-organisms can be killed by heat. This discovery led not only to the sterilization and preservation of foods but to heat-sterilization of surgical instruments.

Antiseptic surgery

The English surgeon Joseph Lister (1827–1912) was impressed with Pasteur's work and looked for other ways than heat to kill germs. Lister particularly wanted to prevent wounds and surgical incisions becoming infected with germs which cause gangrene and other diseases which killed up to half the patients given operations at that time.

Lister found that infections of this kind could be greatly reduced by using a fine carbolic acid spray during operations (see drawing opposite), and by applying carbolic acid dressings to wounds.

Although highly successful, Lister's antiseptic spray only destroyed germs in the air and not those on the surgeon's hands, clothing, and instruments, and those breathed from his mouth. William Macewen (1848–1924) found that the unpleasant carbolic acid spray could be dispensed with if the operating theatre was ventilated with clean fresh air and the surgeon was provided with a sterilized face mask and heat-sterilized instruments.

Macewen's routine gradually evolved into modern aseptic (germ-free) surgery: in addition to sterile face masks and instruments, modern surgeons wear sterile gloves, caps, and gowns; use sterile dressings; and work in filtered germ-free air.

Medical drugs

For centuries people have known of substances which can be used in the treatment of diseases. But it was not until the invention of powerful microscopes that it was possible to study the effects of various chemicals on living germs. The object of this work was to find chemicals which kill germs but not people. The use of these chemicals as medicines is known as **chemotherapy**.

The German chemist Paul Ehrlich (1845–1915) was one of the founders of chemotherapy. He tested hundreds of chemicals before finding one, which he called Salvarsan, that would kill syphilis germs without harming human tissues.

A great step forward in chemotherapy took place in 1935 with the development of a group of chemicals called **sulphonamides**. These affect many different germs including pneumonia diplococci and others which can be fatal.

In 1928 the British microbiologist Alexander Fleming accidentally discovered a chemical produced by one micro-organism which killed other micro-organisms. Chemicals of this type are now called **antibiotics**.

A dish in which Fleming was growing staphylococcus bacteria was accidentally contaminated with a *Penicillium* mould and, to his surprise, Fleming discovered that the bacteria were being destroyed by a substance coming from the mould (see photograph opposite). This substance, later called penicillin, remained unused until 1940 when it was isolated and purified by Howard Florey and Ernst Chain. It was then found very effective in treating pneumonia, meningitis, gonorrhoea, syphilis, and several other infections. Other well-known antibiotics are streptomycin, actinomycin, and tetracycline.

It is now possible to make a naturally produced antibiotic more effective and easier to use by altering its molecular structure. There is now, for example, a partly altered penicillin which, unlike natural penicillin, is unaffected by gastric juice so that it can be taken by mouth.

Resistance to drugs

When a drug is used to combat an infection it kills countless millions of the germs responsible. But, unfortunately, there is sometimes a small number of germs which are resistant and survive the drug treatment. These resistant germs multiply and soon cause a disease which the drug no longer cures. Furthermore, drug-resistant germs can pass their resistance to other, different germs so that the drug loses its effectiveness as a treatment for a wide range of illnesses.

Sensitivity to drugs

People are said to be sensitive to a drug if their body reacts unfavourably to it, developing symptoms such as skin rash or high temperature. It is very important, therefore, that all medicines are taken under medical supervision so that complications like these can be avoided.

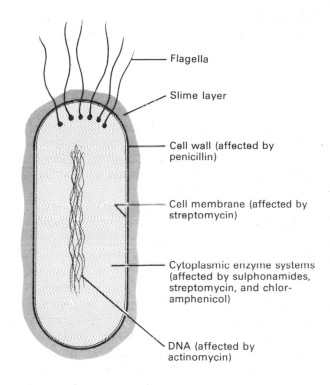

- Flagella
- Slime layer
- Cell wall (affected by penicillin)
- Cell membrane (affected by streptomycin)
- Cytoplasmic enzyme systems (affected by sulphonamides, streptomycin, and chloramphenicol)
- DNA (affected by actinomycin)

Fig. 1. How medical drugs affect bacteria. This diagram indicates the region of a bacterial cell which each drug is thought to affect.

Compare the operating theatre of a hundred years ago (above) with its modern equivalent (below). Why is a patient less likely to be infected with diseases like gangrene in a modern theatre?

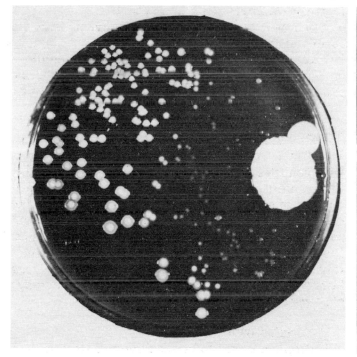

The actual dish in which Alexander Fleming discovered the antibiotic effects of *penicillium* mould. The mould colony is on the right and is surrounded by dying staphylococcus colonies.

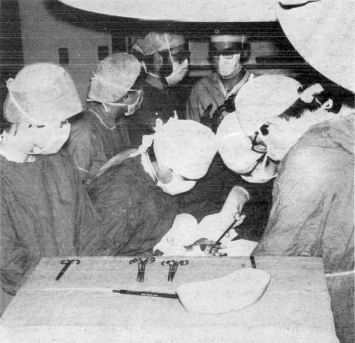

The body possesses many natural defences against infection. These defences include the skin, the membranes covering the eyes and lining the respiratory system, secretions such as tears and digestive juices, and a type of white blood cell called a phagocyte. These are all examples of natural immunity.

The skin

The dead, flattened cells which form the outer layer of the skin form a physical barrier to the entry of germs into the body. As fast as these cells wear away they are replaced by more which grow from below, as described in Unit 43. The same cell-growth repairs the skin when it is damaged.

This barrier of dead cells is kept supple and waterproof by an oily substance called **sebum** produced by the sebaceous glands (Fig. 1C). Sebum also contains an antiseptic chemical which kills some bacteria as they touch the skin.

The eyes

The eyes are protected from infection by a thin skin called the **conjunctiva**. This is continually bathed in an antiseptic liquid produced by the tear glands (Fig. 1A). Blinking spreads this liquid and washes away germs and dust from the eyes.

The respiratory system

The nasal passages, wind-pipe, and bronchial tubes of the lungs are lined with a carpet of microscopic hair-like structures called **cilia** (Fig. 1B). Cells between the cilia produce a sticky fluid called **mucus** which traps germs and dirt breathed in through the nose and mouth. Back-and-forth movements of cilia carries the mucus and trapped germs and dirt to the back of the throat where they are swallowed, rendered harmless by digestive juices, and passed out of the body in the faeces.

The digestive system

Many germs are unavoidably swallowed with food and drink. These germs are usually harmless but in any case the majority are killed by stomach acid (Fig. 1E).

Phagocytes

These are white blood cells which destroy bacteria that invade the body, by engulfing and digesting them. **Neurophils** and **monocytes** are examples of phagocytic cells (Fig. 1D).

Phagocytes are particularly active in wounds. Soon after the skin is damaged capillaries in surrounding tissues dilate, increasing the supply of blood to the area. Fluid pours out of these capillaries into the wound, together with thousands of phagocytes which destroy any bacteria present. Unit 36 describes how the wound is quickly sealed off by a blood clot.

Bacteria which penetrate deep into the body and enter the lymphatic system are killed by large phagocytes inside the lymph nodes, described in Unit 34. Bacteria inside the body are also killed by chemicals called antibodies, described in the next Unit.

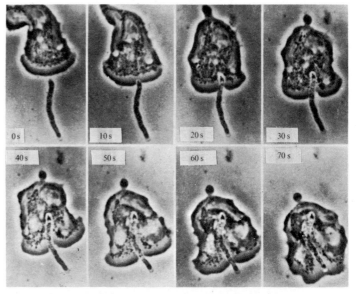

White blood cells (phagocytes) destroying a bacterium.

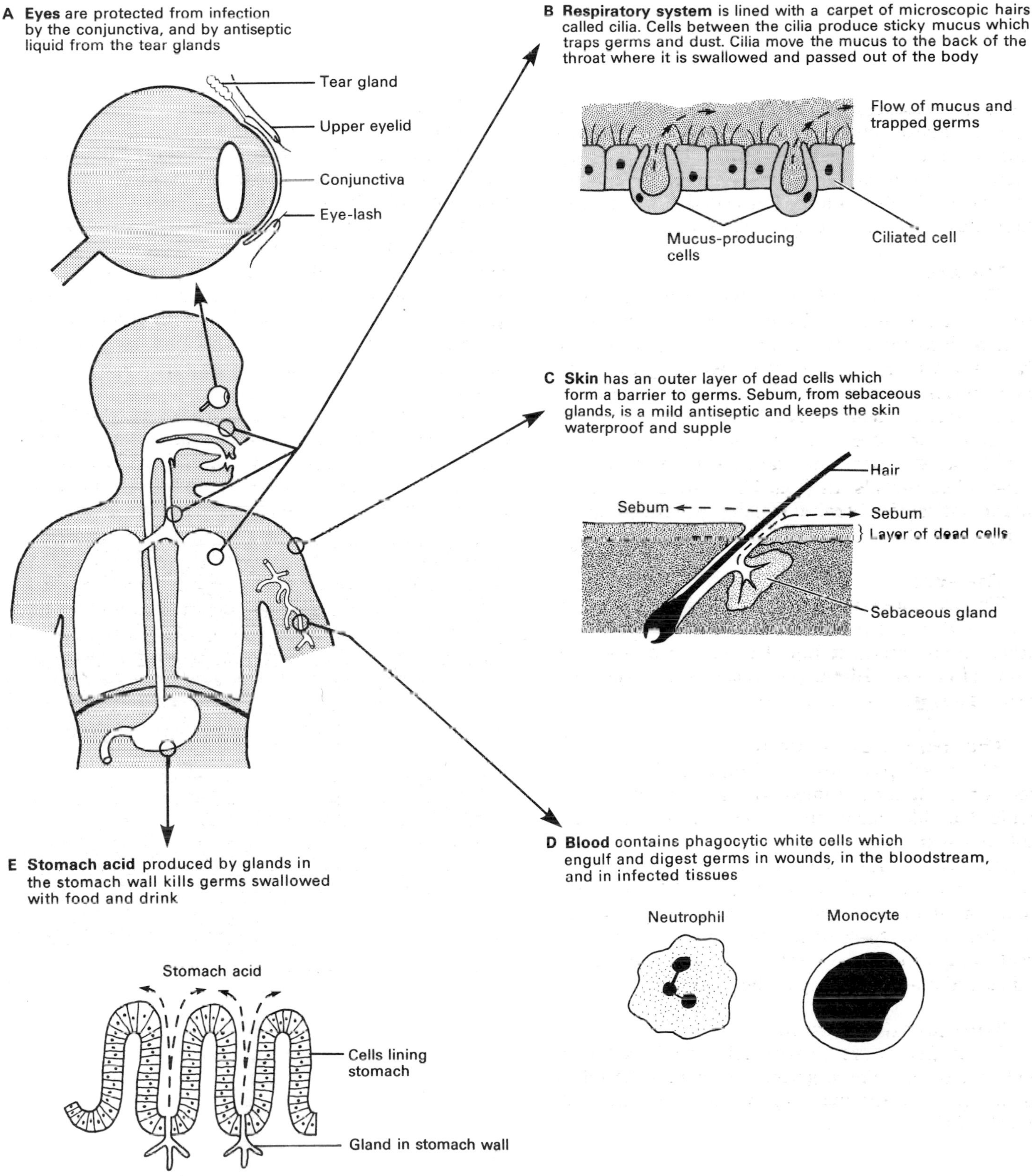

A Eyes are protected from infection by the conjunctiva, and by antiseptic liquid from the tear glands

Tear gland
Upper eyelid
Conjunctiva
Eye-lash

B Respiratory system is lined with a carpet of microscopic hairs called cilia. Cells between the cilia produce sticky mucus which traps germs and dust. Cilia move the mucus to the back of the throat where it is swallowed and passed out of the body

Flow of mucus and trapped germs

Mucus-producing cells

Ciliated cell

C Skin has an outer layer of dead cells which form a barrier to germs. Sebum, from sebaceous glands, is a mild antiseptic and keeps the skin waterproof and supple

Hair
Sebum ← – – – – → Sebum
} Layer of dead cells

Sebaceous gland

D Blood contains phagocytic white cells which engulf and digest germs in wounds, in the bloodstream, and in infected tissues

Neutrophil   Monocyte

E Stomach acid produced by glands in the stomach wall kills germs swallowed with food and drink

Stomach acid

Cells lining stomach

Gland in stomach wall

Fig. 1. The body's natural defences against infection.

Acquired immunity and immunization

When germs penetrate the body's natural defences described in Unit 90, the body reacts by producing substances called **antibodies**. Antibodies circulate in the blood and tissue fluid killing germs or making them harmless. Antibodies also neutralize poisonous chemicals called **toxins** which germs produce.

Since antibodies appear in the body *after* an invasion by germs, their production is an example of **acquired immunity**. There are two types of acquired immunity: active and passive. Antibody formation is an example of **active immunity** because their formation is an active response to infection. Passive immunity is described later in this Unit.

Antibodies

Antibodies are proteins. They are made by white blood cells called lymphocytes, described in Units 29 and 34. Any substance which stimulates lymphocytes to make antibodies is called an **antigen**. Bacteria and viruses are covered with antigen molecules and toxins may also act as antigens.

When lymphocytes contact germs or toxins the antigen molecules are detected and antibodies are formed. Antibodies combine with the antigens producing a number of different effects.

1. **Opsonins** are antibodies which combine with antigen material on the outer surface of germs and appear to make the germs more likely to be destroyed by phagocytes (Fig. 1A). It is as if the antibody makes a germ more 'appetizing'.

2. **Lysins** are antibodies which kill germs by causing them to burst open into fragments which are engulfed by phagocytes (Fig. 1B).

3. **Agglutinins** are antibodies which cause germs to stick together in clumps (agglutinate). In this state the germs can neither penetrate cells nor reproduce properly (Fig. 1C).

4. **Anti-toxins** are antibodies which combine with toxins and render them harmless.

There are many different antigens and each requires a specific antibody to destroy it. The antibody which combines with measles virus antigen, for example, will destroy this virus and no other.

Antibodies are produced slowly at first but if a disease persists for a few days they are produced in much larger quantities. Moreover, the antibodies against certain germs remain in the body for many years after the infection has disappeared, thereby giving the body a built-in resistance to the germs if re-infection occurs. Should this happen, antibodies are produced much faster than during the first infection. A child who has recovered from chicken-pox, for instance, is unlikely to suffer from the disease again despite repeated exposure to infection.

Immunization

The body can be artificially stimulated into producing antibodies. This prepares it in advance to fight off infection. This is done by inoculating someone with **vaccine**. A vaccine is a liquid containing antigens powerful enough to stimulate antibody formation without causing harm. This is called **immunization** because it makes the body immune to germs with the same antigens as the vaccine.

One of the first people to use a vaccine with success was Edward Jenner (1749–1823). Jenner was investigating a theory that people who recovered from a mild disease called cowpox would, thereafter, be immune to smallpox, which is usually fatal. He scratched the skin of a healthy boy and rubbed pus from the hand of a girl with cowpox into the wound. The boy caught cowpox and when he had recovered Jenner inoculated him with pus from a smallpox victim. The boy did not catch smallpox. The modern explanation for this result is that cowpox and smallpox viruses have the same antigens. Thus if someone catches cowpox first they form antibodies which make them immune to smallpox. Jenner's method of immunization soon acquired the name **vaccination** after *vaccinia*, the Latin word for cowpox.

Modern vaccines contain either killed germs, germs made harmless to humans by growing them in non-human hosts, or germ-toxins made harmless in various ways. Vaccines are now available to combat many diseases including typhoid fever, poliomyelitis, cholera, and bubonic plague.

Passive immunity

It is now possible to inject people with ready-made antibodies in order to help them fight infection before their own antibody production has reached full speed. This treatment is called passive immunity: the body is protected without doing any work.

Tetanus antibodies, for example, are obtained by injecting horses with tetanus toxin. The horse makes antibodies to combat the toxin. Some blood is then removed from the horse, and serum, complete with tetanus antibodies, is extracted from it.

A Opsonins are antibodies which combine with antigens on germs, making the germs more likely to be destroyed by phagocytes

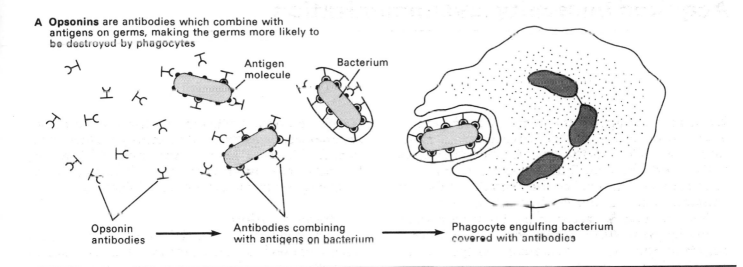

Antigen molecule

Bacterium

Opsonin antibodies → Antibodies combining with antigens on bacterium → Phagocyte engulfing bacterium covered with antibodies

B Lysins are antibodies which break up germs into tiny fragments

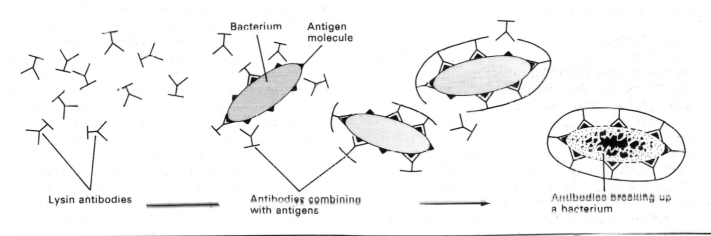

Bacterium   Antigen molecule

Lysin antibodies ——→ Antibodies combining with antigens ——→ Antibodies breaking up a bacterium

C Agglutinins stick germs together so they cannot enter cells or reproduce

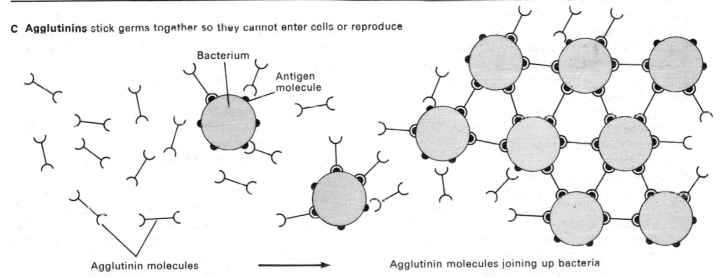

Bacterium

Antigen molecule

Agglutinin molecules ——→ Agglutinin molecules joining up bacteria

Fig. 1. Diagram of how antibodies destroy germs.
Antibodies are made by white blood cells called lymphocytes to protect the body against germs. Each antibody combines with a specific antigen molecule on the germ surface. (Antibody molecules do not actually look like this. These shapes are used merely to indicate antibody–antigen reactions. Moreover, antibody molecules have been drawn much larger than life. In fact, on this scale they would be invisible.

183

Revision tests

1 Humans compared with other animals (Unit 1)
2 Cells (Units 2 and 3)
3 Metabolism (Units 4, 5, and 6)
4 Tissues and organs (Units 7, 8, 9, 10, and 11)
5 The skeleton (Units 12, 13, and 14)
6 Muscles and movement (Units 15, 16, 17, and 18)
7 Photosynthesis and food chains (Unit 19)
8 Types of food and diet (Units 20, 21, and 22)
9 Teeth, digestive system, and the liver (Units 24, 25, 26, 27, and 28)
10 Blood and its functions (Unit 29)
11 The heart and circulatory system (Units 30, 31, 32, and 33)
12 Tissue fluid, blood groups, and blood clotting (Units 34, 35, and 36)
13 The respiratory system, breathing, and gaseous exchange (Units 37, 38, 39, and 40)
14 Homeostasis and excretion (Units 41 and 42)
15 Skin and temperature control (Units 43 and 44)
16 Senses in the skin, tongue, and nose; and sense of vision (Units 45, 46, 47, and 48)
17 Ears, hearing, and balance (Units 49 and 50)
18 The nervous system (Units 51, 52, and 53)
19 The endocrine system (Unit 54)
20 Male and female reproductive systems and menstruation (Units 55, 56, and 57)
21 From fertilization to birth (Units 58, 59, 60, and 61)
22 From birth to old age (Units 62, 63, 64, and 65)
23 Variation, mitosis, and meiosis (Units 66, 67, and 68)
24 Genetics and the genetic code (Units 69, 70, 71, and 72)
25 Microbes and disease (Units 73 and 74)
26 Roundworms, tapeworms, insects, and disease (Units 75, 76, 77, and 78)
27 The spread and prevention of infection (Units 79 and 80)
28 Pure water; sewage and refuse disposal (Units 81 and 82)
29 Other health hazards (Units 83, 84, 85, 86, and 87)
30 The National Health Service and modern medicine (Units 88 and 89)
31 Immunity (Units 90 and 91)

Notes on the tests

Read these instructions carefully before attempting the tests.

1 Answers to most questions can be found by studying the text and illustrations in the units listed at the beginning of each test.

2 Some questions must be answered by drawing conclusions from information given in the text and drawings.

3 *Vocabulary tests* These consist of a list of scientific words and a series of numbered phrases which describe the words in the list. Write out the numbered phrases and then, opposite each phrase, write the word or words which it describes. (*Note*: Many of the words are described more than once, but in different ways.)

4 *Comprehension tests* These must be answered as fully as possible. Most of them, however, require only a paragraph or two.

Test 1 Humans compared with other animals
(Unit 1)

Comprehension test

1 What features do humans, horses, kangaroos, and dolphins have in common?
2 What features do monkeys and humans have in common?
3 In what ways does the body of a human being differ from that of a monkey? (Figures 1 and 3 of Unit 1 will help you answer this question.)
4 a) Anthropoids are social animals. What does this mean?
   b) How does the formation of social groups help monkeys and humans survive?
   c) Describe some of the reasons, other than survival, why humans form social groups.
5 What are the advantages of:
   a) a flat face with forward-facing eyes;
   b) the ability to pull faces;
   c) a large cerebrum; and
   d) hands which can grasp objects?

Test 2 Cells
(Units 2 and 3)

Vocabulary test

nucleus, cell membrane, lysosome, DNA, chloroplasts, chromosomes, RNA, inclusions, chlorophyll, cytoplasm, cells, organelles, photosynthesis, multicellular, endoplasmic reticulum, mitochondria, Golgi body, cellulose, vacuoles, protoplasm, unicellular, ribosomes

1 Organisms which consist of one cell.
2 The part of a cell which contains chromosomes.
3 Absorbs the light energy used for photosynthesis.
4 Units of living matter.
5 Organelles containing enzymes which dissolve dead cells.
6 A semi-permeable membrane.
7 Ribonucleic acid.
8 The process by which a plant manufactures sugar.
9 A word which describes *Amoeba* and *Paramecium*.
10 Only visible during cell division.
11 Organelles which contain RNA.
12 Organisms made of many cells.
13 The complex living material of which cells are made.
14 Plant cells are enclosed in a layer of this substance.
15 The power plants of a cell.
16 Glycogen granules and oil droplets are examples.
17 Concentrates useful substances before they are secreted.
18 All the living material in a cell except the nucleus.
19 The green substance found in some plant cells.
20 Contain enzymes concerned with respiration.
21 The extremely thin 'skin' around a cell.
22 Tubular or slit-like passageways through the cytoplasm.
23 Deoxyribonucleic acid.

24 Mitochondria and ribosomes are examples.
25 In plant cells they are large, permanent spaces in the cytoplasm filled with cell sap.

Comprehension test

1 In what ways are plant and animal cells the same, and in what ways are they different?
2 a) What is the main difference between inclusions and organelles?
   b) Describe the functions of two inclusions and two organelles.
3 Why is a cell membrane described as semi-permeable?
4 What processes in a cell would stop if its nucleus were removed?

Test 3 Metabolism
(Units 4, 5, and 6)

Vocabulary test

enzymes, respiration, ATP, anaerobic respiration, metabolism, oxygen debt, aerobic respiration, anabolism

1 An example of catabolism.
2 Respiration which does not require oxygen.
3 Adenosine triphosphate.
4 Catalysts.
5 Chemical reactions which use energy to build large molecules from smaller ones.
6 The chemical breakdown of food to produce energy for life.
7 Protease is an example.
8 The opposite of catabolism.
9 Occurs in muscle tissue during strenuous exercise.
10 Chemical reactions concerned with synthesis.
11 All the physical and chemical processes necessary for life.
12 Respiration in which oxygen is an essential requirement.
13 Substances which alter the speed of chemical changes without themselves being used up in the changes.
14 Mitochondria contain enzymes which control this process.

Comprehension test

1 a) What is respiration?
   b) Describe some of the ways in which energy released during respiration is used in the body.
2 a) Explain the differences between aerobic and anaerobic respiration.
   b) What is the connection between anaerobic respiration and the formation of an oxygen debt?
   c) How is an oxygen debt 'paid'?
3 'Metabolism is partly synthesis and partly breakdown of molecules.' Explain this statement.
4 Describe the importance of enzymes to metabolism.
5 a) What is meant by enzyme specificity?
   b) What effects do temperature and pH level have on enzyme-controlled reactions?
6 Describe the part played by ATP in transferring energy released by respiration to processes in the body which require energy.

Test 4　Tissues and organs
(Units 7, 8, 9, 10, and 11)

Vocabulary test

*synapse, voluntary muscle, dendron, ligament,
involuntary muscle, cartilage, thorax, axon, cilia,
epithelium, gland, cardiac muscle, neurone, tendon,
myelin, organ, threshold*

1　Muscle which cannot be contracted at will.
2　A sheet of cells covering an external or internal surface of the body.
3　Muscle found only in the heart.
4　Nerve fibre which carries impulses away from the cell body of a neurone.
5　Holds bones together at a joint.
6　Muscle controlled by the will.
7　Microscopic hair-like structures protruding from certain epithelial cells.
8　The chest region of the body.
9　Made of elastic fibres.
10　Muscle attached to the skeleton.
11　Nerve cell.
12　Gristle.
13　Connects muscles to bones.
14　Forms an insulating sheath around nerve fibres.
15　Secretes useful substances such as enzymes.
16　Nerve fibre which carries impulses towards the cell body of a neurone.
17　Made of collagen fibres.
18　The heart is an example.
19　A body cavity which contains the lungs and heart.
20　Stimulation must reach this level before sense organs begin sending out impulses.
21　Occurs at the ends of bones where they rub together.
22　Nerve impulses must cross this gap as they travel from one neurone to the next.
23　This type of muscle is responsible for peristalsis.

Comprehension test

1　a) Define: tissue, organ, organ system.
　　b) Name two examples of each of these.
2　Where are cilia found in the body and what are their functions?
3　Describe the main types of epithelial tissue.
4　a) What features do all muscle tissues have in common?
　　b) Compare the structures of cardiac, involuntary, and voluntary muscle.
　　c) Compare the ways in which these types of muscle contract and their functions in the body.
5　a) What is the main function of nervous tissue?
　　b) How is the structure of nerve cells related to this function?
6　a) What is a nerve impulse?
　　b) What difference will there be between the number of impulses travelling along the optic nerve when an eye is exposed to dim light and to bright light?
7　What is a synapse and how do nerve impulses cross one?

8　a) What do all connective tissues have in common?
　　b) Where are the following connective tissues found in the body, and what is the function of each: areolar tissue, tendons, and ligaments?
9　What are the differences between cartilage and bone?

Test 5　The skeleton
(Units 12, 13, and 14)

Vocabulary test

*periosteum, cranium, axial skeleton, ossification,
diaphysis, appendicular skeleton, marrow, epiphysis,
fixed joints, intervertebral discs, slightly movable joints,
spongy bone, orbits, compact bone, synovial joints*

1　Consists of the skull, vertebral column, and rib cage.
2　Sockets in the skull which protect the eyes.
3　Manufactures blood cells.
4　Consists of the pectoral and pelvic girdles and the arms and legs.
5　Contains and protects the brain.
6　Scientific name for the shaft of a long bone.
7　The process by which cartilage becomes bone.
8　Joints between bones in the roof of the skull are examples.
9　A layer of fibres and cells which encase a bone.
10　Act as shock absorbers in the backbone during running.
11　Joints between the vertebrae are examples.
12　The rounded end or head of a bone.
13　Hard bone which gives the skeleton its strength.
14　The skeleton of a developing embryo undergoes this process.
15　Full of holes but not soft.
16　The knee and shoulder joints are examples.

Comprehension test

1　What are the main functions of: the vertebral column, skull, and rib cage?
2　a) Describe the parts of a long bone such as the femur.
　　b) Explain the functions of each part of this bone, including the marrow.
3　a) What is ossification?
　　b) Write a brief summary of the ossification of a long bone.
4　Name the parts of a synovial joint and explain their functions.
5　a) Name a hinge joint, a ball and socket joint, and a pivot joint.
　　b) How is the shape of the bones at each of these joints related to the movements which the joint can perform?

Test 6　Muscles and movement
(Units 15, 16, 17, and 18)

Vocabulary test

*muscle tone, origin, fracture, insertion, antagonistic
system, flexor muscle, dislocation, sprain, extensor
muscle*

1 A muscle which bends a limb.
2 Separation of bones at a joint.
3 A broken bone.
4 A muscle which straightens a limb.
5 Over-stretched or torn ligaments.
6 The biceps is an example.
7 The fixed anchorage point of a muscle.
8 The triceps is an example.
9 A set of muscles which move a joint in opposite directions.
10 Results from slight tension in the muscles.
11 The triceps and biceps form such a system.
12 The end of a muscle which moves during muscular contraction.

Comprehension test

1 a) Name a joint in the human body. Name the flexor and extensor muscles which move it.
   b) Describe the antagonistic muscle system which moves this joint using the following words in your description: flexor muscle, extensor muscle, origin, insertion, effort, fulcrum, load.
2 a) In which parts of the body are sprains, dislocations, and fractures most likely to occur?
   b) How would you treat each of these conditions?
   c) What must you avoid doing when treating each of these conditions?
3 What is muscle tone and what are its functions?
4 a) Describe a good standing posture and a good sitting posture.
   b) What is the main advantage of maintaining a good posture?
5 a) Describe a bad standing posture and a bad sitting posture.
   b) What harm is done by prolonged bad posture?
6 a) How can exercise and dieting help correct bad posture?
   b) Explain why dim lighting and low work surfaces can lead to bad posture.
7 What is done to the feet by tight shoes and high-heeled shoes?

Test 7 Photosynthesis and food chains
(Unit 19)

Comprehension test

1 'All the food which animals eat comes directly or indirectly from plants.' Explain this statement.
2 a) Define each of the following words: producers, consumers, scavengers, decomposers, herbivores, and carnivores.
   b) Name an example of each of these terms, and explain what part each plays in the formation of food chains.
3 Below are lists of animals and plants which belong to a food chain. Arrange each list into the correct order:
   a) caterpillar, hawk, cabbage leaves, blue-tit;
   b) eagle, stoat, grass, rabbit;
   c) water snail, large water beetle, pond weed duck;

d) microscopic water plants, dragonfly larva, humans, trout, tadpole;
e) earthworm, humans, decaying leaves, chickens;
f) sparrow, sugar, housefly, spider.

Test 8 Types of food and diet
(Units 20, 21, and 22)

Comprehension test

1 Make a copy of chart 1 (see p. 188). Put a tick under one or more of the headings at the right-hand side of the chart to show what the foods in the left-hand column contain.
2 Make a copy of chart 2 (see p. 188). Put a tick under one or more of the headings at the right-hand side of the chart to show which vitamins are described in the left-hand column.
3 Below is a list of foods followed by the number of minutes a person could walk using the energy which each supplies:
   100 g beef   —44 minutes
   100 g bread —48 minutes
   100 g honey—57 minutes
   100 g butter—158 minutes
   a) What conclusions can you draw from these facts?
   b) Which of these foods would supply energy the quickest? (Give reasons for your answer.)
   c) Which of these foods is not normally used by the body to supply energy?
   d) What is the food you have named in (c) normally used for in the body?
4 a) What is meant by the term 'a balanced diet'?
   b) Using the foods listed in Table 1 of Unit 20, design balanced meals for one day to provide about 12 600 kJ of energy.

Chart 1

	Carbohydrates	Fats and oils	Proteins	Vitamins and minerals
1. Baked potato				
2. Fried potato				
3. White bread and butter				
4. Brown bread and butter				
5. Steamed haddock				
6. Fried cod				
7. Orange juice				
8. Raw egg				
9. Ice cream				
10. Egg, lettuce, and tomato salad				
11. Fried pork chop with mashed potato				
12. Cod-liver oil				
13. Cheese and tomato sandwich				
14. Fried bacon and egg				
15. Toffee				
16. Suet pudding				
17. Strawberries and cream				
18. Fried fish, chips, and peas				
19. Milk chocolate				
21. Milk				
22. Fried rice				
23. Steamed rice				
24. Rice pudding				
25. Cream cake				

Chart 2

	A	B_1	B_2	B_{12}	C	D	K
Found in liver							
Found in green vegetables							
Found in oranges and lemons							
Found in yeast							
Found in fish-liver oils							
Lack of it causes rickets							
Found in wholemeal bread							
Destroyed by cooking							
Found in fish							
Lack of it causes beri-beri							
Helps the body obtain energy from food							
Lack of it causes pernicious anaemia							
Needed to heal wounds							
Lack of it causes poor night vision							
Lack of it causes scurvy							
Lack of it may cause bones to bend							
Needed for good vision in dim light							
Made by bacteria in the digestive system							
Helps the body to form protein and fat							
Found in nuts, peas, and beans							
Lack of it causes paralysis of the limbs							
Needed for healthy bones and teeth							
Destroyed by mincing or grating food							

Test 9 Teeth, digestive system, and the liver
(Units 24, 25, 26, 27, and 28)

Vocabulary test

assimilation, gall bladder, ileum, digestion, peristalsis, lipases, crown, absorption, hepatic portal vein, pulp cavity, proteases, faeces, sphincter, villi, amylases, duodenum, deamination, oesophagus, bolus, enamel, large intestine, bile, dentine

1 Forms the hard outer surface of the crown of a tooth.
2 Enzymes which digest proteins.
3 Indigestible material egested from the body.
4 Soft, bone-like substance in a tooth.
5 The process which makes food soluble.
6 Made in the liver.
7 The part of a tooth above the gum.
8 Breakdown of excess amino acids in the liver.
9 The two regions of the gut where villi are present.
10 A ring of muscle.
11 Enzymes which digest starchy food.
12 The space in the centre of a tooth.
13 The part of the gut where water and salts are removed from indigestible material.
14 Muscular contractions which move food along the gut.
15 Tube along which food travels from the mouth to the stomach.
16 Changes fats and oils into an emulsion.
17 Carries blood saturated with dissolved food from the ileum to the liver.
18 Ball of food.
19 Finger-like projections from the intestine wall.
20 Movement of digested food through the intestine wall into the blood.
21 Enzymes which digest fats and oils.
22 Bile is stored here.
23 The scientific word describing what happens to digested food when it reaches the cells of the body.

Comprehension test

1 Study Figure 3 of Unit 25 before answering these questions.
   a) How many milk teeth does a six-year-old child have?
   b) Study the shapes of the milk teeth. Which of the four types of teeth illustrated in Figure 2 of Unit 25 is missing from the milk set?
   c) Which type of milk tooth is the first to be replaced by permanent teeth?
2 Why are two sets of teeth (i.e. milk and permanent) necessary?
3 What is the cause of tooth decay and how can it be prevented?
4 Why is it important to chew food thoroughly?
5 Where do digestion and absorption take place in the body?

6 What has happened to the following foods by the time they are absorbed into the bloodstream: butter, table sugar (sucrose), bread, egg white, glucose?
7 What part do peristalsis and sphincters play in digestion?
8 A villus contains capillaries, a lacteal, involuntary muscle fibres, nervous tissue, and is covered by cells whose exposed surface is folded into microvilli. What is the function of each of these?
9 Explain briefly:
   a) How do the liver and pancreas control the level of glucose in the blood, and why is it important that glucose level remains constant?
   b) How are excess amino acids dealt with by the liver?
   c) How does the liver indirectly prevent loss of blood from wounds?
   d) How does the liver help keep the body warm?
   e) How does the liver aid the digestion of fats and oils?

Test 10 Blood and its functions
(Unit 29)

Vocabulary test

lymphocytes, red cells, leucocytes, plasma, erythrocytes, white cells, haemoglobin, phagocyte, platelets, blood clot

1 Disc-shaped cells which are concave on both sides.
2 Most of these have a large, lobed nucleus.
3 Enables red cells to transport oxygen.
4 The liquid part of blood.
5 Most of them move and change shape like an amoeba.
6 Consists of water and important dissolved substances.
7 Destroyed in the liver and spleen after circulating in the blood for four months.
8 The substance which gives red cells their colour.
9 Another name for red blood cells.
10 Blood which has been changed to a jelly.
11 A word which means 'cell eater'.
12 Cells without a nucleus.
13 White cells made in the lymphatic tissue.
14 Blocks wounds and prevents bleeding.
15 Fragments of cells which are made in the bone marrow.
16 Another name for white blood cells.
17 Slowly releases oxygen to body cells.
18 Their function is to help blood clot in wounds.

Comprehension test

1 List the substances transported around the body by plasma.
2 a) How do red cells and neutrophils differ in appearance, size, and functions?
3 a) Name the types of white cells which are phagocytic.
   b) How do phagocytic cells help fight infection?
4 a) What are platelets?
   b) Describe two ways in which the body would be in danger if there were no platelets.

Test 11   The heart and circulatory system
(Units 30, 31, 32, and 33)

Vocabulary test

cardiac muscle, atria, tricuspid, systole, auricles, arteries, diastole, ventricles, bicuspid, pulmonary artery, cardiac cycle, semi-lunar, veins, pulmonary vein, capillaries, vena cava, aorta

1   The main artery of the body.
2   Valve between the left atrium and ventricle.
3   Technical name for relaxation of the heart.
4   Another name for atria.
5   The heart is made of it.
6   The main vein.
7   Vessels which carry blood away from the heart.
8   Thick-walled chambers of the heart.
9   Carries blood from the heart to the lungs.
10   Takes about 0·8 seconds to complete.
11   Valves situated at the points where blood flows out of the heart.
12   Vessels which carry blood towards the heart.
13   Two names for the upper chambers of the heart.
14   Carries blood from the lungs to the heart.
15   The two lower chambers of the heart.
16   Prevent blood re-entering the ventricles after it has been pumped out of the heart.
17   Technical name for one complete heart beat.
18   Valve between the right atrium and ventricle.
19   Technical name for contraction of the heart.
20   Very narrow blood vessels.

Comprehension test

1   Describe in your own words the events which take place in the heart during one heart beat.
2   a) What is the function of heart valves?
b) What are the differences between semi-lunar valves and the bicuspid valve?
3   a) List the differences between arteries, capillaries, and veins.
b) What is the function of precapillary sphincters?
4   What is the pulse and how is it formed?
5   How does the fact that certain large veins are situated inside muscles help the blood to circulate?
6   a) What is meant by 'double circulation'?
b) Compare what happens to blood as it passes through the pulmonary circulation and as it passes through the systemic circulation.
7   Trace the path of a molecule of sugar from the capillaries of the intestine to the brain, then trace the path of a molecule of carbon dioxide from the brain to the lungs.

Test 12   Tissue fluid, blood groups, and blood clotting
(Units 34, 35, and 36)

Vocabulary test

tissue fluid, lymphatic system, lymph nodes, agglutination, blood donors, transfusion, anti-A plasma, anti-B plasma, haemophilia, fibrin, AB blood group, platelets, heparin, O blood group

1   Returns tissue fluid to the bloodstream.
2   Agglutinates Group B red cells.
3   Two types of antibody.
4   The liquid which transports food and oxygen from the bloodstream to the cells.
5   Transfer of blood from one person to another.
6   Compatible with Group B red cells.
7   A chemical which prevents blood from clotting inside undamaged blood vessels.
8   Glands with narrow channels which have white cells attached to their walls.
9   Provide blood for transfusions.
10   Agglutinates Group A red cells.
11   Red cells sticking together.
12   People with this blood group are known as universal donors.
13   Disorder in which blood clots very slowly.
14   They release thrombokinase.
15   People with this blood group are called universal recipients.
16   Protein threads which help blood clot in wounds.
17   Male children can inherit it from their mothers.

Comprehension test

1   a) What is a capillary bed?
b) How are tissue fluid and lymph formed in a capillary bed?
c) What are the functions of tissue fluid, lymph, and lymph nodes?
2   Citric acid reacts chemically with calcium so that it can no longer react with other substances in blood. In view of this, why is calcium citrate added immediately to blood taken from a donor?
3   a) Describe in your own words what happens to blood in a wound from the moment the skin is cut until bleeding stops.
b) What are the two main functions of the blood-clotting mechanism?
4   Why would it be very dangerous to transfuse Group B blood into a recipient with Group A blood, but quite safe to transfuse Group O blood into the same patient?

Test 13 The respiratory system, breathing, and gaseous exchange
(Units 37, 38, 39, and 40)

Vocabulary test

thoracic cavity, trachea, diaphragm, larynx, bronchi, oxygenated blood, alveoli, bronchioles, inspiration, residual air, expiration, tidal air, pleural membrane, oxyhaemoglobin, vital capacity, deoxygenated blood, intercostal muscles

1 The scientific name for the wind-pipe.
2 Microscopic air sacs.
3 Breathing out.
4 The greatest volume of air which can be breathed in and out of the lungs.
5 Contains vocal cords.
6 The two branches of the trachea.
7 The amount of air which cannot be exhaled from the lungs.
8 Blood without oxygen.
9 Each is surrounded by a network of capillaries.
10 The scientific name for the voice box.
11 Blood which has absorbed oxygen.
12 A dome-shaped sheet of muscle.
13 Narrow branches of the bronchi.
14 Breathing in.
15 The space in the chest which contains the lungs.
16 The volume of air which moves in and out of the lungs during normal breathing.
17 Muscles between the ribs.
18 The red cells of the blood in the pulmonary vein contain this substance.
19 A shiny slippery skin which covers the lungs and inner surface of the thoracic cavity.

Comprehension test

1 Explain, in your own words, how the lungs are inflated and deflated.
2 What important things happen to air as it passes through the nasal passages?
3 What is gaseous exchange and where does it occur in the body?
4 What are the functions of:
   a) the strips of cartilage in the walls of the trachea;
   b) cilia in the air passages to the lungs;
   c) the pleural membrane and pleural fluid?
5 A man runs to catch a bus, then takes a seat and reads a newspaper.
   a) What happens to his breathing rate as he runs, and after he takes his seat?
   b) Describe how the man's respiratory centre controls his breathing rate during these activities.
6 a) What are the differences between blood in the pulmonary artery and blood in the pulmonary vein?
   b) Account for these differences by explaining what happens to blood as it passes through capillaries in the lungs.
7 Study the Table in Unit 40. What conclusions can you draw from the figures given in this table?

Test 14 Homeostasis and excretion
(Units 41 and 42)

Vocabulary test

homeostasis, excretion, glomerulus, osmoregulation, urea, glomerular filtrate, ureters, urine, Bowman's capsule, urinary system, reabsorption, micturition, bladder

1 A liquid which gathers in the bladder.
2 Produced when blood is filtered in a Bowman's capsule.
3 Temporary store for urine.
4 A waste substance produced by the liver.
5 A ball of inter-twined capillaries.
6 A process which changes glomerular filtrate into urine.
7 Removal from the body of waste and unwanted substances.
8 Liquid excreted by the kidneys.
9 Release of urine from the bladder.
10 Contains urea plus many useful substances.
11 Each contains a glomerulus.
12 Removal of useful substances from glomerular filtrate.
13 Carry urine from the kidneys to the bladder.
14 Consists of the kidneys, ureters, and bladder.
15 Maintenance of a constant internal environment.
16 Blood is filtered through its walls into a Bowman's capsule.
17 Regulation of water and dissolved substances in blood and tissue fluid.
18 Scientific name for urination.

Comprehension test

1 a) What is meant by 'homeostasis' and the 'internal environment'?
   b) List the organs concerned with homeostasis and, very briefly, list the homeostatic functions of these organs.
2 a) Define excretion and secretion.
   b) List the substances excreted from the body and name the organs which excrete them.
3 Where is urea produced and why is its removal from the body called nitrogenous excretion?
4 a) Describe the formation of glomerular filtrate.
   b) What are the main differences between glomerular filtrate and urine? and
   c) how do you account for these differences?
5 a) What is osmoregulation?
   b) What is the connection between reabsorption in the kidneys and osmoregulation?

Test 15 Skin and temperature control
(Units 43 and 44)

Vocabulary test

homoiothermic, heat stroke, keratin, sweat glands, poikilothermic, vasodilation, dermis, sebum, germinative layer, vasoconstriction, granular layer, sebaceous gland, hair follicles, dermal papillae, cornified layer

1   Enlargement of blood vessels.
2   A word which describes animals whose temperature depends upon their surroundings.
3   The uppermost layer of the epidermis.
4   Adults have about two million.
5   A word which describes animals which can maintain a constant body temperature.
6   Contraction of blood vessels.
7   Cell division in this layer produces the other layers of the epidermis.
8   Makes cells of the epidermis hard and dry.
9   The layer of the skin below the epidermis.
10 Produce liquid which cools the body.
11 Produce sebum.
12 A mass of blood capillaries just below the epidermis.
13 Occurs in skin blood vessels when body temperature falls below normal.
14 Constantly worn away but replaced by growth from lower layers.
15 Cells in this layer die as they fill with keratin.
16 Hairs grow from these.
17 Occurs in skin blood vessels when body temperature rises above normal.
18 A word which describes temperature control in birds and mammals.
19 Caused by failure of the ability to sweat.
20 Liquid which keeps the skin supple and waterproof.

Comprehension test

1   Define the words homoiothermic and poikilothermic.
2   a) Why does strenuous exercise cause a rise in body temperature?
    b) How is this excess heat removed from the body?
3   Describe the involuntary and voluntary ways in which a constant body temperature is maintained when someone leaves a warm room and goes out-of-doors in cold weather.
4   Why do we feel cooler on a hot dry windy day than on a hot humid windless day even if the air temperature is the same on both occasions?
5   a) Describe the most important ways in which people who fail to wash regularly are endangering their health.
    b) Why is it particularly important to wash hands and nails regularly, to wash between the toes, and to wash combs and hair brushes regularly?

Test 16 Senses in the skin, tongue, nose and sense of vision
(Units 45, 46, 47, and 48)

Vocabulary test

orbits, conjunctiva, extrinsic muscles, sclerotic, cornea, choroid, ciliary muscles, suspensory ligaments, iris, pupil, retina, aqueous and vitreous humour, accommodation, fovea, rods and cones, blind spot, optic nerve, stereoscopic vision, presbyopia, myopia, hypermetropia, astigmatism

1   Made up of nerve endings sensitive to light.
2   A disc made of muscles with a hole at its centre.
3   Hold the lens in place.
4   Prevents light from being reflected around inside the eyeball.
5   Scientific name for short sight.
6   The part of the retina which is not sensitive to light.
7   Transparent substances which fill the eyeball.
8   Transparent skin which covers the front of the eyeball.
9   Conveys nerve impulses from the eyes to the brain.
10 Scientific name for long sight.
11 Has a small diameter in bright light, but is much wider in dim light.
12 A layer of tough fibres in the eyeball.
13 Muscles which control the shape of the lens.
14 The area of the retina responsible for colour vision.
15 Cavities in the skull which contain and protect the eyes.
16 Hold the eyeball in place and move it within its orbit.
17 Light-sensitive receptors.
18 Depends upon having two eyes which can look at the same object.
19 Old sight.
20 Refocusing the eyes between near and distant objects.
21 People with this disorder are unable to focus simultaneously on vertical and horizontal lines.

Comprehension test

1   Describe all that you could discover about your surroundings using only your sense of touch.
2   In what ways do our senses of taste and smell protect us from harm?
3   What is the shape of the lens when the eye is focused on a near object, and on a distant object? How is the lens made to form these two shapes?
4   How are the eyes protected by the eyelids, tear glands, orbits, and the blink reflex?
5   What are the functions of the choroid layer, cornea, suspensory ligaments, and optic nerve?
6   What are the causes of myopia, hypermetropia, and astigmatism? How can these disorders be corrected?
7   a) What differences are there between the fovea and the remainder of the retina?
    b) Why do we lose the ability to distinguish between colours as daylight fades?

8 a) What difference is there between the position on the head of human eyes and the eyes of a horse?
   b) How will this difference affect the field of vision of these two animals?
   c) Which of the two is capable of stereoscopic vision? What are the advantages of this type of vision?

Test 17   Ears, hearing, and balance
(Units 49 and 50)

Vocabulary test

pinna, ear drum, outer ear, Eustachian tube, middle ear, ear ossicles, oval window, inner ear, perilymph and endolymph, semi circular canals, cupula, ampulla, otoliths, cochlea

1 Chain of three bones.
2 Two of them are upright and one is horizontal.
3 Made up of the pinna, ear canal, and ear drum.
4 Its funnel shape directs sounds towards the ear drum.
5 They lever the oval window in and out.
6 A series of tubular passages in the skull bones filled with perilymph.
7 An air-filled space behind the ear drum.
8 This collects sound waves travelling through the air.
9 Tiny pieces of chalk embedded in a jelly-like substance.
10 A swelling at one end of a semi-circular canal.
11 Two types of liquid found in the tubes which comprise the inner ear.
12 This opens during swallowing, letting air in or out of the middle ear.
13 Coiled like the shell of a snail.
14 A cone-shaped lump of jelly.
15 A sheet of skin and muscle which vibrates when sound waves reach it.
16 As vibrations move through it they stimulate sensory nerve endings which send impulses to the brain.
17 Contain sensory nerve endings embedded in a cupula.
18 This ensures that air pressure is equal on both sides of the ear drum.
19 The ear ossicles lever it in and out, which causes vibrations to pass through to the inner ear.
20 Tube connecting the middle ear with the back of the mouth.

Comprehension test

1 Describe what happens to sound waves as they pass through the outer, middle, and inner ear.
2 a) What are Eustachian tubes and what is their function?
   b) How may hearing be affected if the Eustachian tubes become blocked?
3 a) What is the function of the semi-circular canals?
   b) What happens in the semi-circular canals when there is a sudden change in the direction of movement?
4 a) What is the function of the utricles and saccules?
   b) What happens in the utricles and saccules if the speed at which the body is moving changes?

Test 18   The nervous system
(Units 51, 52, and 53)

Vocabulary test

co-ordination, central nervous system, spinal cord, nerve fibres, cerebrum, nerve impulses, cerebellum, medulla oblongata, sensory neurones, reflex action, motor neurones, synapse, peripheral nervous system, meninges, cerebro-spinal fluid, sympathetic stimulation, conditioning, parasympathetic stimulation

1 Thread-like extensions of nerve cells.
2 The part of the brain which controls conscious behaviour.
3 'Messages' which travel along nerves.
4 Nerve cells which convey impulses away from the central nervous system.
5 Situated inside a channel in the backbone.
6 Made up of two cerebral hemispheres.
7 Receives impulses from the organs of balance.
8 A tiny gap between nerve cells.
9 Co-ordinates muscles during walking, dancing, etc.
10 Nerve cells which convey impulses towards the central nervous system.
11 Its outer layer is called the cerebral cortex.
12 A word which describes the parts of the body working together in an orderly way.
13 Prepares the body for action.
14 Made up of all the nerve fibres outside the central nervous system.
15 The process by which a reflex is changed.
16 Made up of the brain and spinal cord.
17 Prepares the body for rest.
18 A liquid which surrounds the central nervous system and fills its hollow interior.
19 Convey impulses from sense organs to the central nervous system.
20 A nerve is made up of thousands of these.
21 An immediate, automatic response to a stimulus.
22 Regulates the rates of breathing and heart-beat.

Comprehension test

1 a) What does co-ordination mean?
   b) Name some examples of co-ordination in the human body.
2 a) What is a reflex action?
   b) Describe the reflex actions which will occur: when dust blows into your eyes; when a bright light suddenly shines in your eyes; when you run fast in hot weather; when food accidentally enters your wind-pipe.
3 What are the functions of the meninges, cerebro-spinal fluid, cerebellum, and medulla oblongata?
4 What is the difference between:
   a) a nerve cell and a nerve;
   b) a sensory neurone and a motor neurone;
   c) white matter and grey matter;
   d) the spinal cord and the spinal column;
   e) the sensory, motor, and association areas of the brain;

f) the cerebral cortex and the cerebral hemispheres;
g) the sympathetic and the parasympathetic nervous systems;
h) unconditioned and conditioned reflexes?

Test 19   The endocrine system
   (Unit 54)

Vocabulary test

pituitary, thyroid, pancreas, testes, adrenals, hormones, target organs, parathyroid, integration, adaptation, insulin, oestrogen, thyroxine, diabetes, testosterone, thyroid-stimulating hormone, negative feed-back, adrenalin, ovaries

1   Glands attached to the kidneys.
2   Parts of the male reproductive system which produce hormones.
3   Produces digestive enzymes and hormones.
4   Gland at the base of the brain.
5   Gland which produces thyroxine.
6   Parts of the female reproductive system which produce hormones.
7   Glands which produce adrenalin.
8   Gland attached to the wind-pipe.
9   These produce testosterone.
10   Gland which produces hormones that control other endocrine glands.
11   Organ which produces insulin.
12   These produce oestrogen.
13   The chemicals produced by endocrine glands.
14   Disease caused by lack of insulin.
15   A female sex hormone.
16   A hormone which prepares the body for action.
17   A hormone which decreases the rate at which the liver releases glucose into the blood.
18   A male sex hormone.
19   Gland embedded in the thyroid gland.
20   Control which involves co-ordination of tissues and organs.
21   Those organs which are affected by a hormone.
22   The name for the way in which the pituitary gland controls other endocrine glands.
23   An organism's response to changes in its surroundings.
24   A hormone which increases the rate at which thyroxine is produced.

Comprehension test

1   What are the main differences between the ways in which hormones and nerve impulses co-ordinate the body?
2   What are giantism and dwarfism and what causes them?
3   What is diabetes and what causes it?
4   What are the secondary sexual characteristics of males and females, and which hormones control them?

5   In what ways does the pituitary gland affect the male and female reproductive systems?
6   Describe the effects of oestrogen on the female reproductive system.
7   Which hormone is produced during sudden emotional stress and what effects does it have on the body?

Test 20   Male and female reproductive systems and menstruation
   (Units 55, 56, and 57)

Vocabulary test

ova, sperms, ovary, testis, puberty, ovulation, scrotum, follicle, oestrogen, fertile, conceive, fallopian tube, uterus, vas deferens, cervix, vagina, vulva, corpus luteum, progesterone, menopause, menstrual cycle, semen, menstruation, epididymis

1   This produces sperms.
2   The age at which sexual maturity is reached.
3   A sequence of events in females which lasts about 28 days.
4   Tube into which ova are released.
5   Scientific name for the womb.
6   Scientific name for the sperm duct.
7   Organ in which a baby develops.
8   This produces ova.
9   Storage area for sperms.
10   The scientific name for the release of an ovum from an ovary.
11   A tube between the uterus and vulva.
12   Bag in which the testes are situated.
13   The liquid which contains sperms.
14   External opening of the female reproductive system.
15   Age at which the production of sex cells begins.
16   Scientific name for eggs.
17   Ring of muscle at the base of the uterus.
18   Begins about 14 days after ovulation.
19   Male sex cells.
20   Female sex cells.
21   To become pregnant.
22   A mass of cells and liquid which enclose a developing ovum inside the ovary.
23   A hormone produced by the corpus luteum.
24   Loss of fertility.
25   A hormone produced by a developing follicle.
26   The part of the follicle left in the ovary after ovulation.
27   A word describing a woman who is able to conceive.
28   A hormone which stimulates the breasts to grow milk-producing tissue.
29   A hormone which stimulates the growth of the uterus lining.

Comprehension test

1   What is an ovum, where do ova develop, and what happens to them immediately after they are released?

2 Describe the events which take place in the female reproductive system during one menstrual cycle.
3 Why is it almost impossible to predict a woman's fertile period?
4 What are the main differences between a sperm and an ovum?

Test 21 From fertilization to birth
(Units 58, 59, 60, and 61)

Vocabulary test

fertilization, embryo, implantation, fertilization membrane, foramen ovale, vernix, pregnancy, placenta, umbilical cord, amnion, foetus, after-birth, zygote

1 Fusion of a sperm with an ovum.
2 The period during which a female carries a developing baby.
3 Protects the baby's skin against amniotic fluid.
4 The bag of liquid which protects a developing baby.
5 The name for an embryo after eight weeks of development.
6 In human females it lasts about nine months.
7 Thick skin around a fertilized ovum.
8 A fertilized ovum.
9 Disc-shaped organ through which a baby obtains nourishment and oxygen.
10 This develops immediately after the ovum is fertilized, preventing entry of more sperms.
11 This contains the artery and vein which connect a baby with its placenta.
12 It is expelled from the womb after a baby has been born.
13 The process by which an embryo becomes embedded in the uterus wall.
14 A hole in the heart between the right and left atria.

Comprehension test

1 Describe fertilization, the functions of the fertilization membrane, and implantation.
2 a) Describe the structure and functions of the placenta.
b) Which organs in the baby's body take over the placenta's functions when the baby is born?
3 What is an amnion and what are its functions?
4 What are the differences between an embryo at week 5, a foetus at week 10, and an 'at-term' baby?
5 a) What are labour pains and what causes them to begin?
b) Describe, briefly, the changes which take place in a baby's circulatory system when it is born. Why are these changes necessary?
c) What is the after-birth?

Test 22 From birth to old age
(Units 62, 63, 64, and 65)

Vocabulary test

colostrum, testosterone, prolactin, puberty, oxytocin, senescence, oestrogen, growth spurt

1 Generally occurs earlier in girls than boys.
2 A hormone which causes the uterus to contract to its usual size after birth of a baby.
3 A liquid produced by the breasts before the appearance of true milk.
4 A baby sucking at the breast stimulates the mother's pituitary gland to release these two hormones.
5 A hormone which stimulates the production of milk.
6 Followed by a period of 'filling-out'.
7 A hormone which causes breasts to eject milk.
8 Ageing.
9 A hormone which stimulates sexual development in boys.
10 A hormone which stimulates sexual development in girls.

Comprehension test

1 Describe the 'primitive reflexes' of a newborn baby. What can a doctor learn by investigating these reflexes?
2 List the advantages and disadvantages of bottle feeding and breast feeding.
3 What are the two advantages of putting a baby to the breast immediately after delivery?
4 Why are regular exercises and baths very important for a mother during the four weeks after delivery?
5 At what age can most babies:
   a) sit unsupported;
   b) stand unsupported;
   c) walk alone;
   d) speak their first words;
   e) reach for and grasp objects;
   f) pick up small objects between thumb and forefinger?
6 What changes must occur in a baby's nervous system before its movements can become fully co-ordinated?
7 List the factors which affect the pattern of a child's growth and development.
8 Which parts of the body:
   a) grow in spurts;
   b) have reached 90 per cent of adult size by the age of six years;
   c) remain small until puberty?
9 At what ages do the main growth spurts occur in boys and girls?
10 Why is the withdrawal method of contraception extremely unreliable?
11 Which contraceptives are best used in conjunction with spermicides?
12 In what way do contraceptive pills prevent conception? Which women should avoid using them?

Test 23 Variation, mitosis, and meiosis
(Units 66, 67, and 68)

Vocabulary test

meiosis, continuous variation, gametes, haploid number, discontinuous variation, homologous chromosomes, inherited characteristics, genetics, X chromosomes, acquired characteristics, Y chromosomes, diploid number, mitosis, genes

1 Sex cells.
2 The full normal number of chromosomes.
3 The scientific study of heredity.
4 Variation in which there are no, or very few, intermediate forms.
5 If a zygote contains two of these it will develop into a female.
6 Cell division which produces daughter cells with the same number of chromosomes as the original cell.
7 Sperms and ova.
8 Characteristics like colour of eyes and hair.
9 Cell division which produces gametes.
10 The number of chromosomes found in gametes.
11 If a sperm containing one of these chromosomes fertilized an ovum the resulting zygote would develop into a male.
12 Reduction division.
13 Characteristics like scars and skills.
14 Intelligence is an example of this type of variation.
15 Chromosomes which form pairs during meiosis.
16 Units of heredity.

Comprehension test

1 What are the differences between: continuous and discontinuous variation; and acquired and inherited characteristics? Give examples of each of these.
2 Why is meiosis called reduction division, and why is it essential for the maintenance of a fixed diploid number of chromosomes?
3 What is the evidence that chromosomes contain genes?
4 What are sex chromosomes?
5 Why are genes called 'units of inheritance'?
6 Describe the main stages of meiosis and mitosis, then list the differences between the two.
7 Where do mitosis and meiosis occur in the body?

Test 24 Genetics and the genetic code
69, 70, 71, and 72)

Vocabulary test

recessive characteristics, DNA, locus, mutation, heterozygous, dominant characteristics, hybrid, phenotype, homozygous, alleles, RNA, sex-linked characteristics, genotype, F_1, monohybrid cross

1 First filial generation.
2 A set of genes which occupy the same locus on homologous chromosomes.
3 A sudden change in a gene or chromosome.

4 Scientific word describing the young from parents who differ in some way.
5 The position of a gene on a chromosome.
6 Scientific word describing a gene pair, the members of which are different (i.e. one dominant and the other recessive).
7 A cross between parents which differ in one way.
8 Can be caused by X-rays.
9 Characteristics more likely to appear in males than females.
10 A characteristic which appears in the young when it is crossed with a contrasting recessive characteristic.
11 Scientific word describing the visible, as opposed to the genetic, characteristics of an organism.
12 Contains the genetic code.
13 Haemophilia and colour blindness are examples.
14 Scientific word describing a gene pair, the members of which are identical.
15 Characteristics which appear in the F_1 generation of a monohybrid cross.
16 The set of genes possessed by an organism.
17 Characteristics which do not appear in the F_1 generation of a monohybrid cross.
18 Controls the types of protein manufactured in cells.
19 Two types of this substance are needed to transcribe the genetic code.

Comprehension test

1 What are pure lines?
2 a) What is a monohybrid cross?
   b) How do the results of a monohybrid cross reveal the existence of dominant and recessive characteristics?
3 Look at Figure 2A of Unit 69.
   a) Which of the parent flies is homozygous dominant and which is homozygous recessive?
   b) In all the F_1 flies, the alleles for wing length are heterozygous and yet these flies have the same phenotype as others in which the alleles are homozygous dominant. Translate this sentence into plain English and explain what it means.
4 Two F_1 *Drosophila* flies were bred together and produced 200 young.
   a) Approximately how many will have the dominant phenotype?
   b) How many will have the recessive phenotype?
5 Pure-line shorthorn cattle with red coats were bred with pure lines having white coats. All their offspring had coats with a mixture of red and white hairs. Use this result to explain what co-dominance means.

Test 25 Microbes and disease
(Units 73 and 74)

Vocabulary test

toxins, pathogens, incubation period, athlete's foot, vectors, kwashiorkor, endemic, parasite, notifiable diseases, epidemic, quarantine, ringworm, pandemic

1 The interval between infection and the appearance of symptoms.
2 Two diseases caused by fungi.
3 An outbreak of disease which spreads through several countries.
4 Animals which carry disease.
5 These diseases must be reported to the Local Health Authority.
6 A sudden outbreak of disease.
7 Parasites which harm their hosts.
8 This word describes a disease which is always present in a particular area.
9 Malarial mosquitoes are examples.
10 Isolation in hospital for a few days longer than the incubation period of a disease.
11 Poisons produced by germs.
12 An organism which lives off another living organism.
13 Smallpox and diphtheria are examples of such diseases.
14 Caused by lack of protein in the diet.

Comprehension test

1 Distinguish carefully between the words pathogen, parasite, germ, infection, and disease.
2 a) List the organisms commonly known as microbes.
   b) Which type of microbe lives *only* as a parasite?
3 List some diseases caused by viruses, bacteria, protozoa, and fungi.
4 What is a secondary infection? List some examples.
5 a) What are the main differences between the structures of a virus and of a bacterium?
   b) Why is it difficult to decide whether or not viruses are alive or dead?
   c) Compare the life-cycle of a virus with that of a spore-forming bacterium.
6 Under favourable conditions bacteria divide in two every 20 minutes. Starting with a single cell, how many bacteria would this rate of division produce in 12 hours?

Test 26 Roundworms, tapeworms, insects, and disease
(Units 75, 76, 77, and 78)

Comprehension test

1 a) Compare the body shape and feeding methods of roundworms and tapeworms.
   b) Which of these two groups live entirely as parasites?
2 a) Make a list of all the features of a tapeworm which enable it to live as an intestinal parasite.
   b) What are the advantages and disadvantages of this way of life?
   c) Why is a tapeworm unable to live outside its host?
3 a) Compare the way in which hookworms and beef tapeworms infect humans.
   b) What precautions must be taken to prevent infection from these parasites?

4 A beef tapeworm produces almost 600 million eggs a year. Why is such a huge egg-production necessary?
5 Which stage of a beef tapeworm's life-cycle is roughly equivalent to the third (infective) larva of a hookworm?
6 What advantage does a flea gain from the shape of its body, and from its ability to jump?
7 a) List the diseases spread by fleas and lice.
   b) How are humans infected with these diseases?
   c) How can the numbers of fleas and lice be controlled?
8 a) List the diseases spread by houseflies.
   b) Describe how the habits of houseflies lead to them spreading the organisms which cause these diseases.
   c) How can the numbers of houseflies be controlled?

Test 27 The spread and prevention of infection
(Units 79 and 80)

Comprehension test

1 List some of the diseases which can be spread by
   a) direct contact with diseased tissue;
   b) coins;
   c) combs;
   d) contaminated food and drink;
   e) dogs;
   f) sneezing.
2 What are healthy carriers and how do they endanger the health of other people?
3 Write an essay with the title 'Personal cleanliness and the prevention of disease'.

Test 28 Pure water; sewage and refuse disposal
(Units 81 and 82)

Comprehension test

1 A sample of water was taken from a deep bore hole, from a reservoir, and from a lowland river where it passed through a city.
   a) Which of these samples is most likely to contain the purest water? Give reasons for your choice.
   b) Which sample is likely to contain the least pure water?
   c) List some of the impurities this least-pure sample could contain and explain where they could have come from.
   d) Describe how impure water can be made pure enough to drink.
2 a) Explain how sewage and refuse can become a hazard to health if they are not disposed of correctly.
   b) Explain the part played by screening, sedimentation, and biological filtration in the treatment of sewage.
3 a) Name some examples of refuse which can be recycled, and explain why this procedure is important.
   b) How are other types of refuse disposed of?

Test 29   Other health hazards
(Units 83, 84, 85, 86, and 87)

Comprehension test

1   Describe some of the ways in which carelessness in the home and bad design of household equipment can result in injury or death.
2   a) Describe some of the differences between air out-of-doors in a country district and air in a poorly ventilated room full of people.
b) Why is lack of ventilation a hazard to health?
3   In an average uninsulated semi-detached house, heat is lost in the following ways:

External walls 35%

Ceilings and roof 25%

Ground floor 15% .

Doors 15%

Windows 10%

How would you go about reducing the heat lost in each of these ways? Use Figure 3 of Unit 84 as a guide.
4   Why is poor lighting at home and at work a health hazard?
5   a) List some of the pollutants which are released into the air and into water from homes, farms, and factories.
b) In what ways can these pollutants endanger human health and food supplies?
6   a) Calculate the cost of smoking 20 cigarettes a day for one year.
b) List the ways in which smoking can make a person unpleasant and even unhealthy to be with.
c) List the diseases which occur more often in smokers than non-smokers.
7   The Health Education Council gives the following advice to those who wish to give up smoking:
a) Cut out the first cigarette of the day to start with, then the second, and so on.
b) Start by cutting out the most 'enjoyable' cigarette of the day, like the one after a big meal.
c) Give up with a friend.
d) Tell everyone you are giving up smoking.
     Thousands of people have found these methods helpful. Try to explain the reasoning behind each piece of advice.
8   Describe some of the ways in which people who drink excessive amounts of alcohol endanger themselves and other people.
9   a) Marijuana is not addictive and does not appear to harm the body. Explain why, despite these facts, it is still considered unwise to take this drug.
b) Draw up a chart showing the 'pleasant' and 'unpleasant' effects of LSD, opiates, barbiturates, and amphetamines.

Test 30   The National Health Service and modern medicine
(Units 88 and 89)

Comprehension test

1   a) How is the National Health Service paid for?
b) What are its aims?
2   a) What services are offered under the heading General Practitioners?
b) Describe the services available to help expectant and nursing mothers and the aged.
3   What precautions are taken in a modern operating theatre to reduce the risk of surgical incisions becoming infected?
4   a) What is meant by chemotherapy?
b) How did the invention of powerful microscopes help in the development of chemotherapy?
5   a) What is an antibiotic?
b) Briefly describe the parts played by Fleming, Florey, and Chain in the development of antibiotics.
6   What is drug resistance and how does it arise?

Test 31   Immunity
(Units 90 and 91)

Comprehension test

1   Describe, very briefly, the differences between
a) natural and acquired immunity;
b) active and passive immunity.
2   What barriers to infection are provided by the eyes, skin, and respiratory and digestive systems?
3   a) What is a phagocyte?
b) Name two examples.
4   a) What is the difference between an antibody and an antigen?
b) Where are antibodies made?
c) Describe how the main types of antibodies function.
5   a) What is a vaccine?
b) What effects do vaccines have when they enter the body?

Glossary

This glossary gives brief definitions of some of the most important scientific words used in the text. A word in *italic* within a definition can be found elsewhere in the glossary.

Absorption   Movement of digested (soluble) food through the walls of the gut into the bloodstream

Accommodation   Refocusing of the eye between near and distant objects.

Acquired characteristics   Characteristics, such as knowledge and scars, which are acquired during a person's lifetime. See *hereditary characteristics*.

Adenosine triphosphate (ATP)   A chemical which transfers energy released by *respiration* to other chemical reactions which absorb energy.

Adrenalin   A *hormone* produced by the adrenal gland. Prepares the body for instant action by increasing the rate of heart-beat, blood pressure, and blood sugar level.

Aerobic respiration   *Respiration* in which oxygen is consumed.

After-birth   Consists of the *placenta* and *umbilical cord* as they are pushed out of the mother's body soon after a baby is born.

Agglutinin   A type of *antibody* which sticks bacteria together, thereby hindering their ability to reproduce.

Agranulocytes   *White blood cells* with clear *cytoplasm*. Most are made in the *lymphatic system*.

Alimentary canal   A tube which extends from the mouth to the anus. Concerned with *digestion* and *absorption*.

Alleles (Allelomorphs)   A set of genes which produce different variations of the same *hereditary characteristic*.

Alveoli   Bubble-like air pockets at the ends of the air passages to the lungs. They are surrounded by *capillaries* and are concerned with *gaseous exchange*.

Amino acids   The chemical units which make up *protein* molecules. These units separate from one another when protein is digested.

Amnion   The liquid-filled space which surrounds and protects a developing *embryo*.

Ampulla   The swollen part of a *semi-circular canal*. Contains sensory hairs and a *cupula*. Detects changes in the direction of movements.

Amylase   A type of *enzyme* which digests starch and glycogen.

Anabolism   Chemical reactions in *cells* in which energy released by *catabolism* is used to build up large molecules from smaller ones, e.g. synthesis of *glycogen* or *proteins*.

Anaerobic respiration   *Respiration* which does not require oxygen.

Antagonistic muscle system   Two sets of muscles which oppose each other at each side of a joint. One set bends the joint and the other straightens it.

Anthropoids   A group of mammals which includes humans, monkeys, gibbons, and apes.

Antibodies   Chemicals made by the body in response to bacteria, viruses, and other *parasites*, the bodies of which contain *antigens*. Antibodies inactivate *antigens* in various ways.

Antigens   Chemicals which stimulate the body to produce *antibodies*. Antigens occur on the bodies of *parasites*, on the surface of *red blood cells*, and on *tissues* foreign to the body.

Antiseptic   A chemical which is used to prevent wounds becoming infected with germs, i.e. *septic*.

Anti-toxins   Chemicals which neutralize the poisons (toxins) which germs release into the body after infection.

Aorta   The main *artery* of the body.

Appendicular skeleton   Parts of the skeleton which are appended (attached) to the *axial skeleton*. Made up of the pectoral and pelvic girdles, and arm and leg bones.

Aqueous humour   Liquid which fills the front chamber of the eyeball (between the lens and *cornea*).

Aquifer   Porous rock filled with water.

Arteries   Blood vessels which carry blood away from the heart.

Arterioles   Branches of *arteries*.

Aseptic   Germ-free.

Assimilation   The process by which *cells* take in and make use of digested food.

Astigmatism   Inability to focus simultaneously on vertical and horizontal lines.

Atlas vertebra   The topmost bone of the *vertebral column*. Permits nodding of the head.

ATP   see *adenosine triphosphate*.

Atria (singular: atrium) (also called Auricles)   Thin-walled upper chambers of the heart. Receive blood from *veins*.

Autonomic nervous system   The parts of the nervous system which control unconscious (involuntary) activities.

Axial skeleton   Parts of the skeleton which form the main axis of the body: the skull, backbone, and rib cage.

Axis vertebra   Bone immediately below the *atlas vertebra* of the *vertebral column*. Permits side-to-side movements of the head.

Axon   The nerve fibre of a *neurone* which conducts impulses away from the cell body of the neurone. See *dendron*.

Balanced diet   The proportions of *carbohydrate* and

protein foods, together with *vitamins* and minerals, necessary to maintain perfect health.

Bicuspid valve   Valve in the heart which prevents blood flowing from the left *ventricle* to the left *atrium*.

Bile   A greenish-yellow liquid made in the liver and passed into the *duodenum*. Helps in the *digestion* of fats and oils.

Binocular vision   Vision possessed by animals (e.g. humans) with two eyes, both of which look at the same object. See *stereoscopic vision*.

Bladder   A bag with muscular walls in which *urine* is stored.

Bladderworm   A stage in the life-cycle of a tapeworm. It lives in a fluid-filled bladder in the muscles of the secondary *host*.

Blind spot   Point at which the optic nerve leaves the *retina* of the eye. It is not sensitive to light.

Bowman's capsules   Microscopic cup-shaped structures in the kidneys concerned with filtration. Each contains a *glomerulus*.

Bronchi   Branches of the wind-pipe.

Bronchioles   Extremely narrow branches of the *bronchi*. End in *alveoli*.

Caecum   A blind-ended branch of the intestine between the *ileum* and *colon*.

Capillaries   Narrow thin-walled blood vessels which carry blood from *arteries* to *veins*. Food and oxygen pass from the blood through capillary walls to the cells.

Carbohydrates   Compounds containing carbon, hydrogen, and oxygen. Include sugary and starchy foods. Carbohydrates are the body's main source of energy.

Cardiac cycle   One complete heart-beat: *systole* and *diastole*.

Cardiac muscle   Muscle which makes up the walls of the heart.

Carnivores   Flesh-eating animals.

Cartilage   Gristle, which occurs at the ends of bones where they rub together (articulation cartilage) and supporting the walls of the wind-pipe. Also forms the skeleton of all vertebrate *embryos*.

Catabolism   Chemical reactions in *cells* in which large molecules are broken down into smaller ones, with the release of energy, e.g. *respiration*.

Cell   A unit of living matter. Consists of a *nucleus*, *cytoplasm*, and a *cell membrane*.

Cell membrane   The *semi-permeable membrane* which forms the outer surface of a *cell*.

Cellulose   A tough material which forms the cell wall surrounding all plant *cells*.

Central nervous system   The brain and spinal cord.

Cerebellum   Region of the brain which co-ordinates muscular activity and controls balance.

Cerebral cortex   Outer layer of the *cerebrum*. Made up of *grey matter*.

Cerebral hemispheres (*cerebrum*)   Two swellings at the top of the brain. Form the largest region of the human brain.

Cerebro-spinal fluid   Fluid which surrounds the *central nervous system* and fills its hollow interior.

Cerebrum   Region of the brain concerned with conscious thought, voluntary movement, and memory. Made up of the *cerebral hemispheres*.

Cervix   A ring of muscle which controls the opening of the *uterus* where it joins the *vagina*.

Chlorophyll   Green substance in plants which absorbs light energy for use in *photosynthesis*.

Chloroplasts   Microscopic objects in plant *cells* which contain *chlorophyll*.

Chondroblasts   *Cells* which manufacture *cartilage*.

Choroid   Layer of black pigment and blood vessels in the eyeball.

Chromosomes   Rod-like objects visible in the *nucleus* of a *cell* during cell division. Contain the hereditary information of the cell in the form of *genes*.

Chyle   Liquid produced in the *ileum* by the action of digestive *enzymes* on food.

Chyme   Semi-liquid, partly digested food resulting from the action of digestive *enzymes* in the *stomach*.

Cilia   Microscopic hair-like structures which project from the surface of certain *cells*. Cilia flick back and forth causing movement of surrounding fluids.

Ciliary body   Circular ridge of blood vessels and *ciliary muscle* fibres inside the eyeball.

Ciliary muscles   Muscles in the eye which change the shape of the lens during focusing.

Cochlea   Spiral tube in the inner ear. Contains sensory nerve endings which respond to sound, producing the sensation of hearing.

Colon   Part of the large intestine. Absorbs water and mineral salts from the *faeces*.

Colostrum   A liquid produced by the breasts before they begin producing milk.

Conditioned reflex   A reflex action which has been changed so that the original involuntary response has been replaced by a response to something different (i.e. to a different *stimulus*).

Condom   A *contraceptive* device consisting of a rubber sheath unrolled over the erect penis before sexual intercourse.

Cones   Light-sensitive receptors in the eye. Operate only in bright light and are sensitive to colour.

Conjunctiva   Transparent skin which covers and protects the front of the eyeball.

Connective tissues   *Tissues* which connect *organs* and other tissues together, protecting and supporting them in various ways, e.g. *tendons*.

Consumers   Organisms in a *food chain* which live by eating (consuming) other organisms in the chain.

Contagious diseases   Diseases spread by direct contact with diseased tissue of another person.

Contraceptives   Devices which prevent conception (sperms fertilizing ova).

Cornea   Transparent circular window at the front of the eyeball.

Corpus luteum   The part of a *follicle* left behind in the ovary after *ovulation*. Produces the *hormone* *progesterone*.

Cranium   The part of the skull which contains and protects the brain.

Crossing Mating together two organisms.

Cupula Cone of jelly attached to sensory hair cells in the *ampulla* of a *semi-circular canal*.

Cytoplasm All the contents of a *cell* except the *nucleus*.

Deamination Breakdown of unwanted *amino acids* in the liver by removal of the nitrogen-containing part of these molecules.

Decomposers Organisms such as certain bacteria and fungi which break down (decompose) dead organisms.

Dendrites Branches of a *dendron*.

Dendron The nerve fibre of a *neurone* which conducts impulses towards the cell body of the neurone. See *dendrites*.

Dentine A substance like bone which forms the inner part of a tooth.

Deoxygenated blood Blood which does not contain oxygen.

Dermis The layer of skin beneath the *epidermis*. Consists of *connective tissue*, blood vessels, nerves, hair roots, and fat.

Diaphragm Dome-shaped sheet of muscle at the base of the *thoracic cavity*. Part of the mechanism which causes breathing in and out.

Diastole Relaxation of the heart during which it fills with blood from *veins*. See *systole*.

Digestion The process by which food is made soluble by the action of digestive juices containing *enzymes*.

Digestive juices Liquid released by *glands* into the digestive system. Contain *enzymes* which break down food into soluble substances.

Diploid number (of chromosomes) Twice the *haploid number*. In most organisms the body cells, but not the *gametes*, contain the diploid number of chromosomes (46 in humans)

Disinfectant A substance which kills pathogenic organisms (germs and other *parasites*) but is also harmful to the human body, e.g. bleach.

Dislocation The separation of bones at a joint so that it can no longer be moved.

DNA (Deoxyribonucleic acid) A chemical found within *chromosomes*. Contains hereditary information, called the *genetic code*, which determines the pattern of development of an organism.

Dominant characteristic A characteristic from one parent which visibly appears in an offspring. The contrasting *recessive* variation of that characteristic (from the other parent) is dominated and is not visible.

Ductless glands *Glands* which make up the *endocrine system*. Produce *hormones*.

Duodenum The region of *alimentary canal* between the *stomach* and *ileum*. Receives *bile* from the liver, and *enzymes* from the pancreas.

Ear drum Sheet of skin and muscle at the bottom of the ear canal. Vibrated by sound waves and transmits these vibrations to the ear *ossicles*.

Effluent The liquid which is discharged from sewage disposal plants.

Embryo An early stage in the development of a baby. Produced by cell division of a fertilized *ovum*.

Enamel Extremely hard white substance which forms the outer surface of a tooth.

Endemic disease A disease which is always present in an area, e.g. malaria in parts of tropical Africa.

Endocrine system A system of *glands* which produce *hormones*.

Endoplasmic reticulum A system of fluid-filled channels within the *cytoplasm* of most *cells*. Probably allows substances to move quickly throughout cells.

Enzymes *Protein* substances which control the rate of chemical reactions in *cells*, generally speeding them up.

Epidemic disease A sudden outbreak of disease which spreads quickly through the population of an area.

Epidermis The outer layer of *cells* in the skin.

Epididymis A long coiled tube which forms a temporary storage area for *sperms* after they pass out of a *testis*.

Epithelial tissues *Tissues* consisting of sheets of *cells* which cover external and internal surfaces of the body, and line the insides of *glands*.

Eustachian tube Tube running from the back of the mouth to the middle ear. Permits air pressure to be equal on either side of the *ear drum*.

Excretion Removal from the body of waste substances produced by *metabolism*, and of substances which are in excess of the body's requirements.

Extensor muscles Muscles which straighten (extend) a joint when they contract.

F_1 generation The first filial generation. The first young produced by *crossing* the parent animals or plants which form the starting point of a genetic experiment.

F_2 generation Young produced by *crossing* or self-fertilizing the members of an F_1 *generation*.

Faeces The indigestible material which remains after *digestion* has taken place.

Fallopian tubes Tubes in the female reproductive system in which *ova* travel from the *ovaries* to the *uterus*.

Fatty acids One set of chemicals released when fats and oils are broken down during *digestion*.

Fertilization Fusion of male and female *sex cells* during reproduction. Results in the formation of a *zygote* which develops into a new organism.

Fibrinogen A *protein* in blood which is converted into fibrin fibres when bleeding occurs. Fibrin forms a blood clot which blocks the damaged vessel and stops bleeding.

Fission Reproduction in many unicellular organisms in which the cell splits into two equal parts (binary fission) or into several parts (multiple fission).

Flexor muscles Muscles which bend (flex) a joint when they contract.

Foetus The stage in the development of a baby at which most of the internal organs have formed and a human appearance has been achieved. Occurs between weeks 8 to 10 of development in humans.

Follicle (Graafian) A fluid-filled space in the *ovary* containing a cell which becomes an *ovum* (egg).

Food chain   A sequence of organisms through which energy is transferred. The chain always begins with *producers* (green plants) and the other links in the chain are *consumers*.

Fovea   Region of the *retina* immediately opposite the lens. Consists of densely packed *cones* and provides the clearest vision.

Fracture   Broken bone.

Gall bladder   A small bladder inside the liver in which *bile* is stored.

Gametes   *Sex cells,* e.g. sperms and ova.

Gaseous exchange   Absorption of oxygen from the air in exchange for carbon dioxide, which is released into the air. Takes place in the *alveoli* of the lungs.

Genes   Units of heredity. The parts of a *chromosome* which control the appearance in an organism of a set of *hereditary characteristics*.

Genetic code   The coded instructions contained in *DNA* molecules which control the development and life processes of an organism.

Genetics   The scientific study of *genes* and the way in which they control *hereditary characteristics*.

Genotype   The genetic make-up of an organism. The set of *genes* which it possesses. See *phenotype*.

Gland   A collection of *cells* which make and release into the body useful substances such as *enzymes* and *hormones*.

Glomerular filtrate   Liquid resulting from the filtration of blood in a *Bowman's capsule*. Contains *urea* and many useful substances.

Glomerulus   A group of *capillaries* inside a *Bowman's capsule* in a kidney. Blood is filtered as it passes through the capillary walls, resulting in *glomerular filtrate*.

Glycerol (glycerine)   One of the substances produced when fats and oils are broken down during *digestion*.

Glycogen   A *carbohydrate* similar to starch. It is stored in the liver and muscles of mammals and is converted into glucose sugar as the body requires energy for *metabolism*.

Goblet cells   *Cells* which produce *mucus*. They are situated in the intestine walls, and between *cilia* cells in the air passages to the lungs.

Golgi body   Flattened spaces enclosed by membranes in the *cytoplasm* of most *cells*. Concerned with *secretion*.

Granulocytes   *White blood cells* with a granular *cytoplasm* and large, lobed *nucleus*. Most are *phagocytes*.

Grey matter   Nervous tissue in the brain and spinal cord which consists of *neurone* cell bodies.

Haemoglobin   Red substance in *red blood cells*. Combines with oxygen and transports it around the body.

Haemophilia   A blood disorder in which the blood clots very slowly.

Hair follicles   Tubular structures in the skin containing the root and lower shaft of hairs. Have *sebaceous glands* attached to them.

Haploid number (of chromosomes)   Half the normal *diploid number* of *chromosomes*. Found in *gametes* as a result of *meiosis*.

Heparin   A chemical in blood and tissues which prevents blood clots forming in undamaged vessels.

Hepatic portal vein   A blood vessel (vein) extending from the small intestine to the liver.

Herbivores   Animals which eat plants.

Hereditary characteristics   Those characteristics which children inherit from their parents, e.g. eye and hair colour.

Hermaphrodite   Bisexual. Organisms which contain both male and female sex organs, e.g. tapeworms.

Heterozygous   An organism is said to be heterozygous for a particular *hereditary characteristic* when the two *genes* which control the characteristic are opposite in nature.

Homeostasis   The maintenance of a constant *internal environment*.

Homoiothermic organisms   Those capable of maintaining a constant body temperature.

Homologous chromosomes   *Chromosomes* which form pairs in the early stages of *meiosis*.

Homozygous   An organism is said to be homozygous for a particular *hereditary characteristic* when the two *genes* which control it are identical.

Hormones   Chemicals produced by *glands* of the *endocrine system*. Help co-ordinate processes such as growth and reproduction.

Host   An organism in, or on which, a *parasite* lives.

Hybrid   An organism which results from *crossing* two organisms which are genetically unlike.

Hypermetropia   Long sight. Inability to focus on near objects. See *myopia*.

Hypothermia   Cooling of the body.

Ileum   The region of *alimentary canal* between the *duodenum* and *colon* where *digestion* is completed and *absorption* takes place.

Immunity   The ability of the body to resist infection.

Implantation   The process by which a developing *embryo* becomes embedded in the *uterus* wall.

Inclusion   Non-living substance such as stored food in the *cytoplasm* of a *cell*.

Incubation period   The period between becoming infected by germs and the appearance of the *symptoms* of disease.

Insecticide   Chemicals used to kill insects.

Insertion (of muscle)   The end of a muscle which moves during muscular contraction. See *origin*.

Insulation   Reduces loss of heat, e.g. the layer of fat beneath the skin is the body's insulation, and in houses insulation consists of material inserted into cavity walls etc.

Insulin   A *hormone* produced by tiny endocrine glands embedded in the *pancreas*. Helps control the amount of sugar in the blood.

Intercostal muscles   Muscles between the ribs which raise and lower the rib cage during breathing.

Internal environment   The environment in which body *cells* live. Formed by *tissue fluid*, which keeps cells warm, supplies their food and oxygen, and removes their waste.

Intra-uterine device (IUD)   A *contraceptive* device inserted into the womb.

Iris   The coloured part of the eye. Consists of muscles which alter the size of the *pupil* and control the amount of light entering the eye.

Iron lung   A mechanical device which makes the chest expand and contract, forcing air in and out of the lungs. Used to help patients unable to breathe normally.

Kwashiorkor   A disease caused by lack of protein in the diet.

Lacteal   A *lymph* vessel which extends through the centre of a *villus*. Absorbs digested fats and oils from the *ileum*.

Large intestine   The region of *alimentary canal* which consists of the *colon* and rectum. Absorbs water and minerals from *faeces* which are then passed out of the body.

Larva   An early stage in the life-cycle of an organism which bears little or no resemblance to the adult, e.g. caterpillar of a butterfly.

Larynx   Voice box. Contains vocal cords.

Leucocytes   See *white blood cells*.

Ligaments   Bands of elastic fibres around a joint in the skeleton. Hold the bones in place.

Lipase   A type of *enzyme* which digests fats and oils.

Lochia   A liquid discharged from the wound in the *uterus* wall caused by removal of the *placenta* following the birth of a baby.

Locus (of a gene)   The position of a *gene* on a *chromosome*.

Lymph   A liquid derived from *tissue fluid* after it has passed between *cells* and drained into the *lymphatic system*.

Lymphatic system   A system of tubes which transport *lymph* from the *tissues* to the bloodstream.

Lymph nodes   Parts of the *lymphatic system* containing *phagocytes* (which remove germs and dead cells from *lymph*) and lymphocytes.

Lymphocytes   *White blood cells* produced in the *lymphatic system*. Produce *antibodies*.

Lysins   *Antibodies* which destroy germs by causing them to burst open.

Lysosomes   *Organelles* containing *enzymes* which dissolve dead *cells*.

Macula (plural: maculae)   A group of *otoliths* in the *utricles* of the inner ear. Concerned with detecting acceleration and deceleration.

Malnutrition   The effects on the body of lack of food or of an unbalanced diet.

Mammary glands   Breasts. *Glands* which produce milk to feed a young baby.

Marrow (bone)   A mass of fat and cells which fills a cavity at the centre of large bones. Manufactures *red* and *white blood cells*.

Medulla oblongata   Region of the brain concerned with regulating blood pressure, body temperature, heart-beat, and breathing.

Meiosis   Cell division which produces *gametes*. Results in *cells* with half the number of *chromosomes* found in the parent cells. See *mitosis*.

Meninges   Membranes which enclose the *central nervous system*.

Menopause   Age at which women lose their ability to have children (between 42 and 55 years of age).

Menstruation   Female period. Breakdown and removal from the body of the lining of the womb. Occurs approximately once a month from *puberty* to *menopause*.

Messenger RNA   *RNA* which becomes an exact copy of part of the *genetic code* and then moves to a *ribosome* where this copy is used by *transfer RNA* to make a *protein* molecule.

Metabolism   All the chemical processes necessary for life.

Microbes   Micro-organisms. Microscopic creatures such as viruses, bacteria, and protozoa.

Micturition   Urination.

Milk teeth   Twenty teeth which are shed between the ages of seven and eleven years.

Mitochondria   *Organelles* concerned with *aerobic respiration* in *cells*.

Mitosis   Cell division resulting in cells with the same number of *chromosomes* as the parent cell. See *meiosis*.

Monohybrid cross   *Crossing* two organisms which possess different variations of one *hereditary characteristic*.

Motor end-plate   The part of a *motor neurone* which is attached to a muscle fibre.

Motor neurone   A *neurone* which conducts impulses from the *central nervous system* to a muscle or a *gland*.

Mucus   A sticky fluid produced by *goblet cells*.

Mutation   A sudden unpredictable change in a *gene* or *chromosome* which alters the way in which it controls development of *hereditary characteristics*.

Myelin   The substance which forms an insulating sheath around nerve fibres.

Myopia   Short sight. Inability to focus on distant objects. See *hypermetropia*.

Nephrons   Microscopic tubules in the kidneys which begin with a *Bowman's capsule*. Concerned with producing *urine*.

Nerve impulse   A wave of electrical and chemical changes which travels along a nerve as a result of a *stimulus*.

Neurones   Nerve *cells*. Consist of a cell body, and nerve fibres which conduct *nerve impulses*.

Notifiable diseases   Diseases which must, by law, be reported to the Local Health Authority as soon as they occur, e.g. smallpox.

Nucleus   The part of a *cell* which contains *chromosomes*. It controls cell *metabolism* and cell division, and contains hereditary material in the form of *genes*.

Nymph   An early stage in the life-cycle of some insects. It resembles the adult except that it is smaller, wingless, and unable to reproduce.

Oesophagus The gullet. A tube extending from the mouth to the *stomach*.

Oestrogen Female sex *hormone*. Controls conditions in the *uterus* before and during *pregnancy*, and controls development during *puberty*.

Opsonins *Antibodies* which combine with chemicals on the surface of germs making the germs more likely to be attacked by *phagocytes*.

Orbits Cavities in the skull which contain and protect the eyes.

Organ A structure made up of several different *tissues* which work together in performing a particular function, e.g. the heart.

Organelles Microscopic objects in the *cytoplasm* of *cells* which take an active part in *metabolism*, e.g. *mitochondria*.

Origin (of muscles) The end of a muscle which remains fixed in position during muscular contraction. See *insertion*.

Osmoregulation Regulation of the flow of water in and out of *cells* by *osmosis*, by regulating the amount of water and/or dissolved material in body fluids.

Osmosis Diffusion of water through a *semi-permeable membrane* from a weak to a strong solution.

Ossicles (of the ear) Three tiny bones in the middle ear which transmit vibrations from the *ear drum* to the inner ear.

Ossification The process by which *cartilage* is transformed into bone during the growth of a baby.

Osteoblasts *Cells* which manufacture bone.

Otoliths Tiny grains of chalk embedded in blobs of jelly attached to sensory hairs in the *utricles* of the inner ear. Displaced by body movements causing hair cells to send impulses to the brain.

Ova (singular: ovum) Female *sex cells* (eggs).

Ovaries Organs which produce female *sex cells* (ova).

Ovulation The release of an ovum (egg) from an ovary.

Oxygenated blood Blood which contains oxygen.

Oxygen debt This occurs in muscle tissue during strenuous exercise when oxygen is consumed faster than it can be supplied by the blood.

Oxyhaemoglobin *Haemoglobin* which has combined with oxygen.

Oxytocin A *hormone* which stimulates the breasts to eject milk, and the *uterus* to contract to its former size after the birth of a baby.

Pancreas A *gland* situated between the *stomach* and the *duodenum*. Produces several digestive *enzymes*.

Pandemic disease A large-scale outbreak of disease which spreads from country to country.

Parasites Organisms which obtain food from the living bodies of other organisms called *hosts*, e.g. tapeworm parasites of humans. *Pathogenic* parasites harm their hosts.

Parasympathetic nervous system The part of the *autonomic nervous system* which prepares the body for rest.

Pasteurization The use of heat for the part *sterilization* of food (especially milk). Named after the French micro-biologist Louis Pasteur.

Pathogenic organisms (pathogens) *Parasites* which harm their *hosts*, e.g. germs.

Pathology The study of *pathogenic organisms*.

Penis The part of the male reproductive system used to transfer *semen* to the female during copulation (sexual intercourse).

Pepsin An *enzyme* produced by the *stomach* which begins the *digestion* of *proteins*.

Periosteum A layer of tough fibres and *cells* which encloses a bone.

Peripheral nervous system Nerve fibres outside the *central nervous system*.

Peristalsis Wave-like contractions of tubular organs such as the gut. Propels the contents of the tube in one direction.

Permanent teeth Teeth which replace *milk teeth*.

Phagocytes *White blood cells* which destroy invading bacteria by engulfing and digesting them.

Phenotype The visible *hereditary characteristics* of an organism as opposed to its *genotype*, or genetic characteristics.

Photosynthesis The process by which plants use light energy trapped by *chlorophyll* to form sugar out of carbon dioxide and water.

Pinna Funnel-shaped outer part of the ear. Collects sound waves.

Pituitary gland The 'master gland' of the *endocrine system*. So-called because it controls most other endocrine organs.

Placenta The organ through which a developing *embryo* obtains food and oxygen from its mother's blood, and passes waste into its mother's blood.

Plasma The liquid part of blood.

Platelets Particles in the blood which take part in blood clotting in wounds.

Pleural membrane A membrane lining the *thoracic cavity* and covering the lungs.

Poikilothermic (organisms) Those which cannot maintain a constant body temperature but vary in temperature according to their surroundings.

Pollutants Substances released by man into the environment where they harm living things.

Pregnancy The period lasting about nine months during which a woman carries a developing baby in her *uterus*.

Presbyopia Old sight. A condition which occurs in old age in which the lens loses its ability to change shape during focusing.

Producers (in a food chain) Green plants which produce food and form the starting point of *food chains*.

Progesterone A *hormone* produced by the *corpus luteum* and *placenta* during *pregnancy*. Stimulates growth of the *uterus* and growth of milk-producing tissue in the breasts.

Proglottids Segments which make up the body of a tapeworm.

Prolactin A *hormone* which stimulates the breasts to produce milk.

Prostate gland A *gland* in the male reproductive system which takes part in the formation of *semen*.

Protease A type of *enzyme* which digests *proteins*.

Proteins The main body-building foods. Meat, eggs, and fish contain proteins.

Protoplasm The living material of a *cell*. Divided into *cytoplasm* and a *nucleus*.

Puberty Sexual maturity. The stage of development at which males and females become able to reproduce.

Pulmonary circulation System of blood vessels carrying blood from the heart to the lungs and back again.

Pulp cavity A cavity in the centre of a tooth containing blood vessels and nerves.

Pupa The stage in the life-cycle of an insect when it stops feeding and moving and changes into an adult insect.

Pupil The hole in the *iris* of the eye which controls the amount of light entering the eye.

Pus Consists of dead cells and germs. Found in infected wounds. See *septic*.

Quarantine Isolation of people who may be infected with a disease. They are isolated for a few days longer than the *incubation period* of the disease.

Reabsorption (in kidneys) Extraction of useful substances from *glomerular filtrate* as it passes along the *nephrons* of a kidney.

Receptors The parts of a sense organ which receive a *stimulus* and convert it into *nerve impulses* which travel to the *central nervous system* along a *sensory neurone*, e.g. the *rods* and *cones* of the eyes.

Recessive characteristic A characteristic of one parent which does not visibly appear in an offspring when it is crossed with another contrasting (*dominant*) variation of that *characteristic*.

Red blood cells Disc-shaped blood *cells* without a *nucleus*. Contain *haemoglobin* and transport oxygen from the lungs to the rest of the body.

Reduction division See *meiosis*.

Reflexes Responses which occur very quickly without conscious thought. See *conditioned reflex*.

Residual air The volume of air which still remains in the lungs even after forcible breathing out.

Respiration A sequence of chemical reactions in *cells* which release energy.

Respiratory centre Part of the brain which controls the rate of breathing.

Retina A layer of light-sensitive *cells* at the back of the eye on which images are projected.

Rhesus factor An *antigen* which occurs on the *red blood cells* of the majority of humans (85 per cent in Britain). Those with the factor are called Rh-positive and those without are Rh-negative.

Ribosomes *Organelles* which are attached to the *endoplasmic reticulum* or float free in the *cytoplasm*. Contain *RNA* and are concerned with *protein* manufacture.

RNA (Ribonucleic acid) A chemical found in the *ribosomes* and *cytoplasm* of *cells*. Concerned with *protein* manufacture.

Rods Light-sensitive receptors in the *retina* of the eye. Function in dim light and are not sensitive to colour.

Saccules Fluid-filled spaces of the inner ears with the same function as the *utricles*.

Saliva Fluid produced and released into the mouth by three pairs of salivary *glands* in response to food. Consists of the *enzyme* salivary *amylase* and mucin.

Scavengers Organisms such as carrion crows and earthworms which eat dead organisms.

Sclerotic Layers of tough white fibres which form the outer wall of the eyeball.

Scolex The head of a tapeworm. Consists of suckers and sometimes bears hooks.

Scrotum Bag of skin which contains the *testes*.

Sebaceous glands *Glands* in the *hair follicles*. Secrete an oily substance called *sebum* which makes skin supple, waterproof, and mildly *antiseptic*.

Sebum Chemical produced by *sebaceous glands*. Keeps the skin supple and waterproof, and mildly *antiseptic*.

Secretion Release by *glands* of useful substances such as *hormones* and *enzymes*.

Semen Fluid produced in the *testes* and other *glands* of the male reproductive system. Consists of *sperms* and chemicals which nourish them and stimulate their swimming movements.

Semi-circular canals Three curved tubes in the inner ear. Contain sense organs which detect changes in the direction of movements.

Semi-lunar valves Valves situated at the point where blood flows out of the heart. Prevent it flowing back into the *ventricles*.

Semi-permeable membrane A membrane which allows certain substances to pass through but prevents the passage of others, e.g. the *cell membrane*.

Senescence Ageing.

Sensory neurone A *neurone* which conducts *nerve impulses* from a sense organ to the *central nervous system*.

Septic Usually refers to wounds which have become inflamed and full of *pus* owing to infection with germs.

Septic tank A large concrete-lined tank sunk in the ground in which sewage is allowed to decompose.

Sex cells *Cells* produced by the reproductive organs. *Sperms* and *ova* (eggs). Fuse together during *fertilization* and form a *zygote*. See *gametes*.

Sex chromosomes *Chromosomes* which determine whether a fertilized egg develops into a male or a female.

Small intestine The *duodenum* and *ileum* of the *alimentary canal*.

Sperm duct A tube which conveys *sperms* from a *testis* to the *penis*.

Spermicides *Contraceptive* chemicals which kill sperms entering the female during sexual intercourse.

Sperms Male *sex cells*.

Sphincter A ring of muscle found in the walls of tubular organs such as the digestive system. Contraction of a sphincter slows or stops movement of substances through the tube.

Spinal cord   A thick bundle of nerve fibres which extends down from the brain through a canal inside the *vertebral column*.

Spleen   An organ immediately below the *stomach* which produces *white blood cells* and destroys old, worn-out *red blood cells*.

Spores   Reproductive cells which give rise directly or indirectly to new organisms. Spores often have protective walls which enable them to survive for long periods in conditions which would kill the fully formed organism.

Sprain   Injury to a joint caused by over-stretching or tearing of *ligaments*.

Stereoscopic vision   Vision which produces a three-dimensional impression of objects looked at. Depends upon *binocular vision*: two eyes looking at the same object from slightly different angles.

Sterilization   Making something completely free of germs; or making an organism incapable of reproduction.

Stimulus   Something which causes a sense organ to send out *nerve impulses*, e.g. light, sound, or touch.

Stomach   A bag-like organ at the end of the *oesophagus*. Begins the digestion of *proteins*.

Stomata   Tiny pores in the leaves and other parts of plants through which water evaporates and *gaseous exchange* takes place.

Stretch receptors   Nerve endings which detect stretching of muscle fibres and other *tissues*.

Suspensory ligaments   Fibres which hold the lens in position within the eye.

Sweat gland   A *gland* in the skin which produces a liquid that evaporates into the air cooling the body.

Sympathetic nervous system   The part of the *autonomic nervous system* which prepares the body for action.

Symptoms (of disease)   The effects of germs and other *parasites* on the body, e.g. headache, vomiting, and pain.

Synapse   A microscopic gap over which *nerve impulses* must pass when moving from one *neurone* to the next.

Synovial joints   Freely movable joints in the skeleton, e.g. knee and elbow.

Systemic circulation   The blood vessels which carry blood from the left *ventricle* of the heart around the body and back to the heart at the right *atrium*.

Systole   Contraction of the heart. Causes it to pump blood into the *arteries*. See *diastole*.

Taste buds   Groups of sensory nerve endings in the tongue which respond to certain chemicals entering the mouth, producing the sensation of taste.

Tendons   Bands of strong fibres which attach muscles to bones.

Testes (singular: testis)   The parts of the male reproductive system which produce *sperms*.

Testosterone   A *hormone* which controls the changes that occur at *puberty* in males.

Thoracic cavity   A cavity in the chest containing the lungs, heart, and main blood vessels.

Threshold   The level of stimulation at which *nerve impulses* being flowing from a sense organ, or begin crossing a *synapse*.

Thrombin   An *enzyme* which takes part in the clotting of blood in wounds.

Thrombokinase   An *enzyme* released from *platelets* and damaged *tissues*. It causes blood to clot in wounds.

Thyroid gland   A *gland* of the *endocrine system*. Produces the *hormone* thyroxin, which influences physical and mental development.

Tidal air   The volume of air which moves in and out of the lungs when at rest.

Tissue fluid   Fluid forced through *capillary* walls into spaces between cells, providing them with food and oxygen and removing their waste products.

Tissues   Groups of similar *cells* specialized to perform a particular function, e.g. muscular and nervous tissues.

Toxins   Poisonous chemicals produced by *pathogenic organisms*.

Trachea   Wind-pipe. A tube extending from the mouth to the lungs.

Transfer RNA   *RNA* which picks up *amino acids* and, in conjunction with *messenger RNA*, builds them up into *protein* molecules.

Transfusion (of blood)   Transferring blood from a blood donor to someone whose life is in danger from loss of blood.

Transpiration   Evaporation of water from plant *cells* and out of pores called *stomata*.

Tricuspid valve   Valve in the heart which prevents blood flowing from the right *ventricle* to the right *atrium*.

Umbilical cord   A tube containing blood vessels which connect a developing *embryo* with its *placenta*.

Urea   A substance produced by the liver from ammonia and excreted by the kidneys in *urine*. Ammonia results from the *deamination* of excess *amino acids*.

Ureters   Two narrow tubes which carry *urine* from the kidneys to the *bladder*.

Urethra   A tube leading from the *bladder* to the outside of the body.

Urinary system   Consists of the kidneys, *ureters*, and *bladder*. Concerned with *excretion*.

Urine   A liquid containing *urea* and various mineral salts. Results from *excretion* by the kidneys.

Uterus   Womb. A bag-like organ of the female reproductive system. Contains, protects, and nourishes a developing baby.

Utricles   Fluid-filled spaces of the inner ears. Contain sense organs called *maculae* which detect changes in the speed of movements.

Vaccine   A suspension of dead, inactivated, or relatively harmless germs which, when introduced into the body, stimulate the production of *antibodies* and make the body immune to attack from certain germs.

Vacuoles   Fluid-filled spaces in the *cytoplasm* of *cells*.

Vagina   The part of the female reproductive system which extends from the cervix to the outside of the body.

Vasectomy   *Sterilization* of men which involves cutting the *sperm ducts*.

Vasoconstriction   Constriction (narrowing) of blood vessels.

Vasodilation   Expansion (dilation) of blood vessels.

Vectors   Organisms which carry *pathogenic microbes* (germs) from one *host* to another.

Veins   Blood vessels which carry blood towards the heart.

Venae cavae   The main *veins* of the body.

Ventricles   Thick-walled lower chambers of the heart. Pump blood into *arteries*.

Vernix   A white grease which covers and protects the skin of a *foetus* after about week 28 of development.

Vertebral column   The backbone, or spine. A chain of small bones called vertebrae which support the body, protect the *spinal cord*, and permit bending movements.

Villi (singular: villus)   Minute finger-like structures on the inner surface of the small intestine. Digested food is absorbed through villi into the bloodstream and *lymph*.

Vital capacity (of lungs)   The greatest volume of air which can be inhaled and then forced out of the lungs.

Vitamins   Organic chemicals required in small amounts in the diet to maintain health.

Vitreous humour   Jelly-like substance which fills and supports the back of the eye.

Vocal cords   Membranes in the *larynx* which vibrate, producing sounds which form the voice.

Vulva   The external opening of the female reproductive system.

White blood cells (leucocytes)   A number of different colourless *cells* in the blood, e.g. *granulocytes, agranulocytes, lymphocytes,* and *phagocytes*.

White matter   Nervous tissue in the brain and spinal cord which consists of nerve fibres

Zygote   The *cell* which results from the fusion of a *sperm* and an *ovum* after *fertilization* has occurred.

Index

Page numbers in **bold** type indicate the main references in the text.

Abdomen 22
Absorption 48, **54**
Accidents
  and alcohol 174
  in and around the home 166
Accommodation (*see* Focusing)
Addiction (*see* Drugs)
Adenosine diphosphate (*see* ADP)
Adenosine triphosphate (*see* ATP)
ADP 12
Adrenalin **106**, 108, 109
After-birth 122
Ageing 128
Agglutination 70
AIDS 148
Air pollution 170
Albinos 140
Alcohol 56, 128, **174**
Alcoholism 174
Alimentary canal (*see* Digestive system)
Alveoli 74, 78, 80
Amino acids 40, 48, 54, **56**, 82, 84, 142, **144**
Amnion 118
Amoeba 4
Amphetamines 174
Ampulla 100
Amylases 48, 52
Anabolism 8
Anaemia 42, 150, 174
Ancylostoma 150
Anthrax 154, 156, 158
Anthropoids 2
Antibiotics 146, **178**
Antibodies 70, **182**
Antigens 70, 182
Antiseptics 178, 182
Anti-toxins 182
Anus 52
Aqueous humour 92
Areolar connective tissues 20
Arteries 60, 62, **64**, 66, 84, 128, 172
Arterioles 64, 86
Artificial respiration 78
Astigmatism 96
Athlete's foot 88, 146, **148**, 158
ATP 12
Atria 60, 62, 122
Autonomic nervous system 102, **106**

Baby (*see also* Embryo and Foetus)
  birth 122, 172
  development **120**, 124, 126, 128
  newborn **124**, 176
Backbone 24, 102
Bacteria 38, 42, 46, 48, 58, 88, 92, 146, **148**, 150, 152, 158, 164, 168, 170, 178, 180, 182
Balance (sense of) 100, 102
Barbiturate drugs 174
Bed bugs 154

Beri-beri 42
Bicuspid valve 60, 62
Bile 52, **56**
Binocular vision 96
Birth 120, **122**
Birth control 130
Bladder **82**, 106
Bladderworm 152
Blind spot 94, **96**
Blood 26, **58**, 66, 86, 108, 114, 120, 150
  circulation of 22, **60–66**, 122
  clotting 42, 58, 72, 140
  groups **70**, 140
  purification in liver 50, **84**
  sucking by insects 154
  transfusion 70
  transport of carbon dioxide 58, 68, 74, 76, 78, **80**, 118
  transport of food 48, 54, 56, 58, 118
  transport of oxygen 58, 66, 74, 76, 78, **80**, 118, 122
Boils 148, 158
Bone 20, **24–28**, 30, 32, 34, 42, 128
Bone marrow 24, **26**, 58
Bottle feeding 124
Bowman's capsules 84
Brain 2, 94, 98, **102**, 104, 109, 120, 170, 174
Breast feeding 124
Breasts 2, 109, 114, 118, **124**
Breathing 36, 74, **76**, **78**, 102, 109
Bronchi 74
Bronchioles 74, **78**, 106
Bronchitis 172
Bubonic plague 154, 158, 182

Calcification (of bone) 26
Calcium 42, 78
Cancer 146, 170, **172**, 174
Capillaries 54, **64**, 66, 74, 84, 86
Carbohydrates 38, **40**, 42, 44, 48, 54
Carbon dioxide
  and gaseous exchange **80**, 82
  from respiration 10
  in photosynthesis 38
  transport by blood 58, 68, 74, 76, 78, **80**, 118
Cardiac muscle 60
Carnivores 38
Cartilage **20**, 26, 74
Catabolism 8
Cells **4–6**, 82, 109
  and genes 134, 144
  and tissues 14–20
  chondroblast 20
  division 134, 136
  fibroblast 20
  macrophage 20
  mast 20
  membrane of 4, 6

  muscle 16
  nerve 18
  ova 109, 112, 114, 116
  red blood 58, 64, 70, **80**
  sperm 109, 110, 116
  structure of 6
  white blood 58, 68, 182
Cellulose 4, 48
Cerebellum 102
Cerebral cortex 102
Cerebral hemispheres 102
Cerebro-spinal fluid 102
Cerebrum 102
Cervix 112, 116, 122
Chain, Ernst 178
Change of life 114
Chicken-pox 148, 158
Chlorophyll 4, **38**
Chloroplasts 4
Cholera 146, 148, 156, 158, 182
Chondroblasts 20
Choroid layer 92
Chromosomes 6, **134–44**
Chyle 54
Cilia **14**, 74, 112, 180
Ciliary body (*see* Ciliary muscles)
Ciliary muscles 92, **94**, 96
Circulatory system **58–68**, 122
Cochlea 98
Collagen fibres 20, 64
Colostrum 124
Colour blindness 140
Colour vision 94
Common cold 148
Conception 112, **116**, 120, 130
Conditioned reflexes 104
Cones 94
Conjunctiva 92, 182
Connective tissue 16
  areolar 20
  bone 20
  cartilage 20
  fibrous 20
Constipation 146
Contagious diseases 158
Contraceptives 130
Co-ordination 36
  development of 126
  hormonal 108–9
  nervous 102–6, 115
Copper poisoning 170
Copulation 110, 116
Cornea 92, 96
Coronary heart disease 146, 172
Corpus luteum 114, 118
Cranium (*see* Skull)
Cretinism 109
Cupula 100
Cyanide poisoning 170

208

Cytoplasm 4, 6, 18

Deamination 56, 82
Decomposers 38, 46
Defecation 48
Deoxyribonucleic acid (see DNA)
Department of Health and Social
    Security 176
Dependence (upon alcohol and drugs) 174
Development
    control by genes 134–44
    of a baby 120, 124, 128
Diabetes 109, 146 ·
Diaphragm 24, 74–6, 78
Diet 44, 128, 146
Digestion 36, 40, 48, 52, 156
Digestive system 42, 48, 50, 52–4, 182
Diphtheria 146, 148, 158
Disease
    and immunity 176, 177, 180–2
    and medicine 178
    carriers 158
    causes of 146, 172, 174
    control of 150, 152, 154, 156
    vitamin deficiency 42
Dislocations (of joints) 32
DNA 6, 140–4, 148
Dominant characteristics 138, 140
Drinking (see Alcohol, and Alcoholism)
Drosophila 138, 140
Drugs 56
    addiction to 174–5
    medical use of 178
    non-medical use of 174–5
    resistance to 178
    sensitivity to 178
Ductless glands (see Endocrine glands)
Duodenum 52
Dwarfism 109
Dysentery 146, 148, 156, 158

Ear drum 98
Ear ossicles 98
Ears 2, 24, 98, 170
Effluent 164
Eggs (ova)
    bed bug 154
    flea 154
    hook worm 150
    human 109, 112, 114, 116, 120
    lice 154
Ehrlich, Paul 178
Elastic fibres 20, 62, 64
Embryos
    human 26, 116, 118, 120
    tapeworm 152
Emphysema 172
Endemic diseases 146
Endocrine glands 58, 108–9, 118
Endoplasmic reticulum 6
Energy
    and food chains 38
    from food 40, 42, 44
    from the liver 56
    from respiration 10, 12
    sunlight 38
Entamoeba 148

Enzymes 6, 8, 12, 14, 48, 52, 54, 142, 152
Epidemic diseases 146
Epididymis 110
Epithelial tissue 14
Eustachian tube 98
Excretion 82–4
Exercise 10, 12, 36, 64, 76, 78, 86, 128
Eyes 2, 24, 42, 92–6, 100, 168, 182

Faeces 48, 52, 164
Fallopian tube 112, 114, 116
Fatigue 36
Fats 38, 40, 42, 44, 48, 52, 54, 56, 109
Fatty acids 48, 54
Fertilization 112, 116, 120, 130, 132, 136,
    138
Fibrinogen 56, 58, 72
Fibrous connective tissue 20, 64
Flatworms (see Tapeworms)
Fleas 88, 146, 154, 158
Fleming, Alexander 178
Florey, Howard 178
Focusing (of eyes) 92, 94, 96
Foetus 120, 122
Food
    and diet 44
    chains 38, 170
    contamination 158
    digestion 48, 52–4, 56
    poisoning 46, 146, 148, 158
    preservation 46, 178
    produced by photosynthesis 38
    transport by blood 48, 54, 56, 58
    types of 20, 21
Food chains 38, 170
Food poisoning 46, 146, 148, 158
Fovea 94
Fractures (of bones) 32
Fungi 38, 46, 88, 146, 148, 158

Gametes (see also Ova and Sperms) 136,
    138, 140
Gamma rays 170
Gangrene 178
Gaseous exchange 80
Gastric juice 52
Genes 134–44
Genetic code 140–4
Genetics 132–44
Genotype 140
German measles 176
Germs 38, 52, 72, 74, 88, 124, 148, 154,
    156, 158, 164, 168, 178, 180–2
Giantism 109
Glands 14, 56, 108, 114, 116
Glomerulus 84
Glucose 10, 38, 48, 54, 56, 82, 84, 106, 109
Glycerol 54
Glycogen 56, 106
Golgi body 6
Gonorrhoea 158
Growth 36, 40, 42, 109, 128

Haemoglobin 42, 58, 80
Haemophilia 72, 140
Haemorrhage 42
Hair 2, 36, 88, 154

Hallucinations 174–5
Health
    and alcohol 128
    and diet 40, 42, 44
    and exercise 36
    and food preservation 46
    and prevention of infection 42, 152, 154,
        158
    and smoking 128, 172
    at home and work 168
    in old age 128
    Local Health Authority 146, 176
    mental 176
    National Health Service 176
    of babies 124, 172
    of nursing mothers 124
    of skin and hair 88
Hearing 98, 102, 170
Heart 16, 22, 36, 60–2, 64, 66, 102, 106,
    109, 120, 122, 146, 176
Heart attacks 146, 172
Heating (of homes) 168
Heat stroke 86
Heparin 72
Hepatic portal vein 54, 56
Herbivores 38
Hereditary characteristics 116, 128, 132,
    134, 138
Heredity 132, 138–44
Heroin 174–5
Homeostasis 82
Homoiothermic 86
Homologous chromosomes 136
Hookworms 150
Hormones 14, 108–9, 114, 118, 124
Houseflies 156, 158, 164
Hybrids 138
Hygiene
    in the home 160
    newborn babies 124
    nursing mothers 124
    personal 160
    skin and hair 88
    teeth 50
Hypermetropia 96
Hypothermia 168

Ileum 52, 54
Immunity 176, 180–2
Immunization 182
Implantation 116, 130
Inclusions 6
Incubation period 146
Infection
    immunity to 176, 180–2
    prevention of 42, 152, 154, 158
    spread of 146, 152, 154, 156, 158, 164, 174
Influenza 146, 148, 158
Inheritance (see Heredity)
Insecticide 154, 156
Insects 88, 146, 154, 156, 158
Insulation (of homes) 168
Insulin 56, 109
Intelligence 132
Intercostal muscles 76, 78
Iodine 42
Iris 92, 106

Iron 42, 56
Iron lung 78

Jenner, Edward 182
Joints **28**, 30, 32, 36, 102

Kidneys 56, 58, 70, 82, **84**, 128
Koch, Robert 178
Kwashiorkor 146

Labour pains 122
Lactic acid 10
Large intestine 52, 148
Larvae
  flea 154
  hookworm 150
  housefly 156
Larynx 74
Lead poisoning 170
Lens 92, **94**, 96
Leprosy 158
Leucocytes **58**, 68
Leukaemia 170
Leverage (in skeleton) 30
Lice 88, 146, **154**
Life histories
  bacteria 148
  bed bug 154
  beef tapeworm 152
  flea 154
  hookworm 150
  housefly 156
  virus 148
Ligaments 20, 28, 32, 34
Lighting (of homes) 168
Lister, Joseph 178
Liver 48, 54, **56**, 58, 82, 86, 106, 109, 174
Local Health Authority 146
Lochia 124
Long sight 96
LSD (lysergic acid diethylamide) 174
Lungs 58, 66, **74–80**, 82, 106, 122, 170, 172
Lymphatic system 68, 180
Lymphocytes 58, **68**, 182
Lymph vessels 54, 68
Lysins 182
Lysosomes 6

Macrophage cells 20
Macula 100
Magnesium 42
Malaria 146, 158
Mammals 2, 86
Mammary glands 2, 109, 114, 118, **124**
Marijuana 174
Mast cells 20
Measles 148, 158
Medulla oblongata 102
Meiosis 134, **136**, 138
Meninges 102
Menopause 114
Menstrual cycle 114
Menstruation **114**, 118, 120
Mercury poisoning 170
Metabolism **8**, 48, 82, 86
Microbes (*see also* Bacteria and Fungi) 146, **148**

Milk 40, 42, 46, 109, 114, 118, **124**, 158
Minerals (in diet) **42**, 44, 54, 56, 86
Mites 88
Mitochondria 6
Mitosis 134
Monohybrid cross 138
Morphine 174
Mosquitoes 146, 158
Motor areas (of brain) 102
Mucus 14, 74, 130, 182
Mumps 148
Musca 156
Muscles 10, 12, 24, 64, 128
  antagonistic 30
  cardiac 16, 60
  co-ordination of 36, **102**, 124, **126**
  diaphragm 24, **74–6**, 78
  fatigue of 36
  flexor and extensor 30
  intercostal 76, 78
  involuntary 16, 64
  movement and leverage of 30
  tone 32, 36
  voluntary 16, 102
Muscle tone 32, 36
Mutation **140**, 142
Myelin 18, 126
Myopia 96

National Health Service 176
Negative feed-back 108
Nematode worms 146, **150**, 156
Nephron 84
Nerve cells (*see* Neurones)
Nerve impulses **18**, 78, 94, 98, 102, 104, 108
Nerves 18, 90, 92, 94, 98, **102**
Nervous system **102**, 106, 128
Nervous tissue **18**, 102
Neurones **18**, 102
Noise (as a pollutant) 170
Notifiable diseases 146
Nucleus 4, **6**, 116, **134–44**

Obesity 128
Oesophagus 52
Oestrogen 109, 114, 118, 128
Old sight 96
Opium 174
Opsonins 182
Optic nerve 92
Organelles 6
Organ systems 22
Osmoregulation 84
Osmosis 56, 84
Ossification 26
Otoliths 100
Ova 109, **112**, 114, **116**, 120, 130, 138
Oval window 98
Ovaries 109, **112**, 114
Ovulation **112**, 118, 130
Oxygen 102
  and gaseous exchange **80**, 82
  and respiration 10
  debt 10
  from photosynthesis **38**, 168
  reduced by pollution 170

transport by blood 58, 66, 74, 76, 78, **80**, 118, 122
Oxytocin 124

Pancreas 52, 56, 82, 109
Pandemic diseases 146
Paramecium 4
Parasites 38, **146**
  bacteria 148
  bed bugs 154
  beef tapeworm 152
  fleas 154
  fungi 148
  lice 154
  nematode worms 150
  protozoa 148
  viruses 148
Pasteur, Louis 46, **178**
Pasteurization 46, **178**
Pathogens (*see* Parasites)
Pectoral girdle 24
Pediculus 154
Pelvic girdle 24
Penicillin 178
Penis **110**, 130
Pepsin 52
Periods (*see* Menstruation)
Periosteum 26
Peristalsis 16, 52, 106, 112
Perspiration 84, **86**
Phagocytes 58, 68, 180
Phenotype 140
Phosphorus 42
Photosynthesis 4, **38**, 170
Phthirus 154
Protozoa 146, **148**, 158, 164
Pituitary gland 109, 114, 118, 124
Placenta 118, 120, 122
Plasma 58, 70, 80
Platelets 58, 72
Platyhelminths (*see* Tapeworms)
Pleural fluid 74
Poikilothermic 86
Pollution 164, **170**
Posture 34, 36
Potassium 42, 56
Pregnancy 44, 112, 114, **118**
Presbyopia 96
Progesterone 114, 118
Prolactin 124
Prostate gland 110
Protease 8, 48
Proteins 38, **40**, 42, 44, 48, 52, 54, **56**, 68, 82, 142, **144**, 146
Protoplasm 4
Puberty 110, 112, **128**
Pulex 154
Pulse 60
Pure lines 138

Quarantine 146

Rabies 148, 158
Radioactive waste 170
Reabsorption (by kidneys) 84
Recessive characteristics **138**, 140
Red blood cells 56, **58**, 64, 70, **80**

Reflex arc 104
Reflexes 104, 124, 172
Refuse disposal 164
Reproductive systems 128
    female 112, 114, 116, 118
    male 110, 116
Respiration (see also Breathing and
    Lungs) 40, 80, 82, 86
    aerobic 10
    anaerobic 10
    and ATP 12
Respiratory centre (of brain) 78
Retina 92, 94, 96
Rhesus factor 70
Ribonucleic acid (see RNA)
Ribosomes 6, 144
Ribs 24, 74–6
Rickets 42
Ringworm 146, 148, 158
RNA 6, 144
Rods 94
Roundworms 146, 150

Saccules 100
Safety (in the home) 166
Saliva 52, 106, 156, 158
Salt (see Sodium chloride)
Sap 4
Scarlet fever 146
Scavengers 38
Sclerotic layer 92
Scrotum 110
Scurvy 42
Sebaceous glands 88, 180–1
Semen 110, 116
Semi-circular canal 100, 102
Semi-lunar valves 60, 62
Seminal vesicle 110
Senses
    balance 100
    hearing 98
    pain 90
    pressure 90
    smell 90, 102, 172
    taste 90, 102, 172
    touch 90, 102
    vision 92–6
Sensory areas (of brain) 102
Serum 182
Sewage disposal 150, 152, 158, 164, 170
Sex cells (see Gametes)
Sex chromosomes 136, 140
Sex hormones 109, 114, 118, 128
Sex-linked inheritance 140
Short sight 96
Skeleton 24, 26, 28, 30
Skin 36, 40, 42, 82, 86, 88, 128, 148, 150,
    154, 180

Skull 24, 92, 100, 102
Sleep 36
Small intestine 52, 54
Smallpox 146, 148, 158, 182
Smell (sense of) 90, 102, 172
Smoke (as a pollutant) 170
Smoking (tobacco) 128, 172
Sodium chloride 42, 46, 90
Spectacles 96
Sperm duct 110
Sperms 109, 110, 116, 130, 138
Sphincters 52, 64, 82, 106
Spleen 58
Spinal cord 24, 102, 104
Spores 148, 158
Sprains 32
Stereoscopic vision 96
Sterilization
    of food 44, 178
    to prevent conception 130
Stomach 52
Stomata 38, 170
Sugar 38, 40, 46, 54, 56, 84, 90
Suspensory ligaments 92, 94
Swallowing 52
Sweat glands 84, 86, 88
Sweating 86, 106
Sympathetic nervous system 106
Synapse 18, 102
Synovial fluid 28
Synovial joint 28
Syphilis 158

Taenia 152
Tapeworms 38, 152, 156, 158
Taste 90, 102
Tear glands 92, 182
Teeth 42, 44, 48, 50
Temperature control 82, 86, 102, 124
Tendons 20, 30
Testes 109, 110
Testosterone 109, 128
Tetanus 182
Thoracic cavity 22, 74, 76
Thorax 22
Thrombin 72
Thyroid gland 109
Thyroxine 109
Tissue fluid 64, 68, 82
Tissues 22, 134
    connective 20
    epithelial 14
    muscular 16
    nervous 18
Tongue 90
Touch 90, 102
Toxins 146, 182
Trachea 52, 74

Tricuspid valve 60, 62
Tuberculosis 156, 158
Typhoid fever 146, 156, 158, 182
Typhus 154, 156

Umbilical cord 118
Urea 56, 58, 82
Ureters 82
Urethra 82, 110
Urinary system 82, 84
Urine 82, 84, 109, 164
Uterus 112, 114, 116, 118, 120, 122, 124
Utricles 100

Vaccination 146, 176, 182
Vacuoles 4
Vagina 112, 122
Variation 132
Varicose veins 64
Vas deferens (see Sperm duct)
Vasoconstriction 86
Vasodilation 86
Vectors 146, 154, 158, 164
Veins 60, 62, 64, 66
Venereal disease 148, 158
Ventilation 168
Ventricles 60, 62
Venules 64
Vernix 120
Vertebrae 24, 28, 102
Vertebral column 24, 28
Villi
    intestinal 54
    placental 118
Viruses 146, 148
Vision 42, 92–6, 102
Vitamins 40, 42, 44, 46, 54, 56, 84
Vitreous humour 92
Vocal cords 74
Vulva 112, 124

Water pollution 170
Water supplies 162
White blood cells 58, 68, 182
Whooping cough 158
Wind-pipe 52, 74
Womb (see Uterus)
Wounds 42, 58, 73, 158, 178, 180

X-chromosomes 136
X-rays 170, 176

Y-chromosomes 136

Zygote 116, 134, 136, 140